NORWEGIAN WOOD

Chopping, Stacking, and Drying Wood the Scandinavian Way

NORWEGIAN

-WOOD-

Chopping, Stacking, and Drying Wood the Scandinavian Way

Lars Mytting

With a Foreword by
Roy Jacobsen

Translated from the Norwegian by
Robert Ferguson

MACLEHOSE PRESS
QUERCUS • LONDON

The scent of fresh wood
is among the last things you will forget
    when the veil falls.
The scent of fresh white wood
in the spring sap time:
as though life itself walked by you,
with dew in its hair.
That sweet and naked smell
kneeling woman-soft and blond
in the silence inside you,
using your bones for
a willow flute.
With the hard frost beneath your tongue
you look for fire to light a word,
and know, mild as southern wind in the mind,
there is still one thing in the world
you can trust.

—Hans Børli

PAGE 2 Great heights of artistry are possible if you combine chunky and finely chopped logs with unsplit branches. This stack was made by Arthur Tørisen, formerly a postmaster at Kvam, in Gudbrandsdal, when he was seventy-six years of age.

OPPOSITE Birch bark is waterproof and has many uses. Here, it is used to protect a traditional Norwegian cord stack from rain—a method used for hundreds of years.

OVERLEAF Earlier, thin sticks of spruce or aspen were called *kitchen wood*. Because it burns fast and intensely the temperature can be easily controlled, and it was the preferred wood for making food in cast-iron kitchen stoves. Thick logs of birch were called *living-room wood*.

CONTENTS

FOREWORD: WOODCHOPPING

In my experience, chopping wood is a personal business. I've often wondered what kind of chopper of wood I am, whether I am the stoic type, like Kjell Askildsen, a fellow Norwegian writer who can stand there with his ax for hours on end and think just *one* thought. Think it all the way through. Or the more sanguine type, who forgets all his cares as the wood chips fly and the woodpile grows. Or maybe more like my father, the slightly neurotic type, the hoarder, typical of the generation of Norwegians who lived through the Second World War, with all its shortages. When my father died we found out that the reason he kept his Mazda parked out in the street was so that he could fill his garage with wood, something in the region of thirty-five to forty cubic meters of it. I inherited the last load in its entirety, drove it home in a truck, and stacked it in the garden, and the cellar, and the shed. Thirteen years later there's still some of it left, even though we burn wood all the time.

Or what about the aesthetic type, the poet, who uses a tape measure to make sure all his work is cut to exactly the same length and exerts himself to produce logs that resemble one another as closely as possible, slender and conical, which he stacks with a military precision at some prominent place on his property, often adding a little timber roof so that the whole thing ends up looking like a piece of sculpture such as you might come across in the pages of a coffee-table book.

How about the standard bungler, often found among that group of young ramblers and environmentalists, skiers and fly-fishers, who seldom build more than a campfire—the law permitting—with the help of a little green toy ax and a little green folding saw bought for an outrageous price from the outdoor-sporting-goods store, the kind of person who thinks he already knows all about wood and so never learns anything about it. These types hold forth with a connoisseur's air about pine twigs and birch bark and decaying birch, which they affect to despise, preferring pine themselves, though they should watch it here: In the eastern parts of Norway, pine is under threat from the elk. They just snort when told that beech makes the best firewood, being the wood that gives the most kilowatt joules per cubic meter, with oak and ash in second and third places and birch competing with rowan a little further down the table; lower still come pine and spruce—and that's ignoring the cherry and apple trees and a number of other hardwoods (not necessarily deciduous trees), not to mention the gray alder, which doesn't warm any more than balsa and cardboard no matter how dry and well seasoned it is, way down in eighth place (I think), which we can safely leave to the beavers, who are, fortunately, making their return to the Norwegian fauna. I've been missing the beavers.

The industrial approach also deserves a mention. Representative exponents are those middle-aged and humorless natives who will occasionally use an ax and a wedge, but much prefer one of the two types of hydraulic splitters, the electrical one and the one you mount on a tractor. They wear orange helmets, insulated vests, safety glasses, hearing protectors, gloves, and steel-toed boots, even though they're standing on level ground in their own backyards in the autumn sunlight, with the cutter rigged up between two old pallets so the logs drop straight down into the trailer, which they will afterward reverse into the old barn and deposit in that empty space over there on the right, where in times past the hay was kept. It's useful work, but it gives them no particular pleasure. The pleasure comes afterward, when they can turn off the machine, get out of all that protective gear, and light up a hand-rolled cigarette.

With the exception of the fumbler and the aesthete, there's probably a little bit of all these characters in me, though I can't help feeling there's something missing from my collection of archetypes. Because to tell the truth, I am starting to get fed up with chopping wood. The sheer amount we go through at the cabin means I've been doing far too much woodcutting recently, so that right now I'm probably an unhappy mixture of the collector and the idiot (the man with the safety glasses and earmuffs), whose mood as he goes about his work mingles annoyance, irritation, and impatience. On top of all that, it's become a bore. And here it now seems obvious to me that I have omitted all mention of the desperado and the melancholic, not to mention the psychopath, representing the darker sides of the human psyche, impossible to ignore in any serious consideration of the art of chopping wood—we are, after all, talking about *smashing something into pieces*. With all the physical power at one's disposal, using a sharpened cutting edge—for thousands of years the most efficient battlefield weapon of them all. Modern life doesn't offer many opportunities to compare with this, to engage in a serious act of violence one day and enjoy the fruits of it the next, and all without doing anyone any harm. Am I a part-time psychopath?

So that is what is most likely to be going through my mind as I address the chopping block these days—the idea that what I'm about to do connects me with history. It reminds me of who I am, and where I come from.

Roy Jacobsen

PREFACE: THE OLD MAN AND THE WOOD

I can still conjure up vividly the day when I realized that a wood fire is about so much more than just heat. It wasn't a cold winter's day. In fact, it was late April. I had put the summer tires on my Volvo weeks earlier, and scraped my skis clean of last year's wax, and I was all ready for the Easter holidays.

We had moved out here to the little town of Elverum, in southeastern Norway, just before Christmas. With the help of a block heater for the car and a couple of fan heaters in the house, we had made it through the last half of a not particularly arduous winter for the Østerdal region. A couple of retirees lived in the house next door: decent people of the generation born before the Second World War, hardworking and cheerful. Ottar, the man of the house, had trouble with his lungs and hadn't ventured outdoors much that winter.

On that particular spring day, with a gentle breeze blowing across the fields and water from the winter's thaw glinting brown in the ditches, nothing was further from my mind than thoughts of the winter now behind us.

A tractor pulling a trailer stopped outside and backed into the neighbors' driveway. Revved up the engine, tilted the trailer—and dumped an enormous pile of birch logs in front of the house.

Enormous? The load was gigantic. You could feel the ground tremble as the logs came thudding down.

Ottar appeared in his doorway, wheezing. He looked tired and unwell. This was a man whose most extensive outing since last November had been the walk down the path to his mailbox by the fence and back up again.

He stood there, studying the birch logs. Then he changed from his house slippers into outdoor shoes, closed the inside door behind him, and headed over toward the pile, navigating his way carefully around the muddy puddles. He bent down and picked up a couple of logs, weighing them in his hands, and began chatting to the farmer, who had by now turned off the engine.

Firewood *now*? I thought. When what everybody else is looking forward to is that first glass of beer out on the veranda?

Sure enough. Now was the time. I learned that later from Ottar. Wood should always be bought in April or May. Unseasoned wood. That way the drying process can be properly controlled, the price is lower, and you can get hold of as much of it as you need.

I stood watching from my kitchen window as the tractor left, and Ottar began to shift the wood.

He began to stack it.

Once more he was able to enjoy the feeling of doing something meaningful.

In the beginning each log seemed to exhaust him and he rested frequently, wheezing and panting for breath. I went over and we exchanged a few words. Thanks anyway, but no, he didn't need help. "Good wood this year," he volunteered. "Feel this one. Or this one. Beautiful white bark. Evenly cut, they've used a well-sharpened chain saw, you can tell from the way the chip here is square. I don't use a saw myself anymore. I'm too old. This has been neatly chopped too. You don't always get that now, not now that everybody's using a wood processor. Anyway, I must get on."

And Ottar went back to work, and I went back inside. Not long afterward I took a drive around the area and I noticed how buying wood in the spring was something that everybody here seemed to do. Especially in front of the older-looking houses: always a woodpile. Stocking up, like buying your ammunition in preparation for the elk-hunting season. Or canned food before you set off on a polar expedition.

A week went by and Ottar's pile of wood wasn't looking any smaller. Not until another week passed did I notice the top of the pile was slightly flatter now. And wasn't there a change in him too? Didn't he seem to have a bit more of a spring in his step?

We began talking. He didn't really have that much to say about what he was doing. Words weren't necessary. For a man who had suffered his way through a long winter, struggling against age and ill health, a man who had once been able-bodied and up to the challenge of any physical labor, here at last was a job where things made sense again. Once more he was able to enjoy the feeling of doing something meaningful, and the sense of calm security that comes to the man who knows he is well prepared, he is early, he has time on his side.

I never tried to get Ottar to talk about his feeling for wood. I preferred to watch him in action, peacefully getting on with the job. It was basically so simple and straightforward and yet, in the way *he* did it, there was also something almost noble about it.

Just once, he mentioned something that was not strictly practical: "The scent is the best thing of all," he said. "The scent of fresh birch. Hans Børli—my favorite poet—wrote a poem about the scent."

Ottar spent a month on his woodpile. Stopping now and then, but never for too long, to savor the smell, and the smell of sap from the smattering of spruce logs that came with the load. Until one day there was nothing left but the twigs, chippings, and bark, which he gathered up for use as kindling.

I've never seen a man change quite the way he did. Old age and infirmity were still there, but with this sudden flowering of spirit and energy he was able to keep them in their place. He started taking short walks, he stood more erect. One day he even powered up a bright yellow lawn tractor and cut the grass.

Was it just the activity and the summer warmth that made him better? I don't think so. It was the wood. All his life he'd chopped his own firewood. And although he'd put away his chain saw for good now, he still enjoyed the feel of each log in his hand, the smell that made him feel he was at work inside a poem, the sense of security in his stack, the pleasing thought of the winter that lay ahead, with all those hours of sitting contentedly in front of his woodburning stove. In much the same way, I suppose, that no one gets tired of carrying bars of gold, he knew that what he held in his hands was his insurance against the cold to come.

That's how this book was born. In my rear-wheel-drive Volvo 240, my quest took me to some of the coldest places in Norway to visit the burners and choppers of wood. I have stopped at crossroads to listen for the buzz of a chain saw or—best of all—the faint creaking sounds of an old man at work with a bow saw. Made my careful approach and tried to bring the conversation around to the subject of wood.

The factual material in this book represents the distilled wisdom of my encounters with people who are passionate about wood, enthusiasts as well as professional researchers. I have benefited greatly from my conversations with experts in the fields of combustion and silviculture. And, not least, from the series of research reports published annually under the modest title *Proceedings of the Norwegian Forest and Landscape Institute*.

Along the way I've tried out most of the techniques I've been introduced to. I've dried finely chopped oak in our kitchen oven, struggled to build a beehive woodpile, miscalculated the trajectory of a felled pine. And I've been on a quest to discover the soul of the wood fire. But wood people don't always like to put their passion into words. This is something you have to discover for yourself, in the tall, elegantly shaped woodpiles, in the fresh layer of caulk applied to an old black woodburning stove, in an open woodshed with its long wall angled south (don't worry, all will be explained later). Thus much of this book is concerned with *method*, because it is about feelings that are communicated through method. On publication it attracted a surprisingly large readership throughout Scandinavia, selling in excess of two hundred thousand copies in Norway and Sweden alone. Firewood enthusiasts from all parts of the world wrote to share their own experiences, and the most useful and important of these have been included in this new edition.

I hope the concentration on method will also make this a useful book, because if it omitted all mention of tree felling, soapstone stoves, how to sharpen chain saws, and different ways of stacking wood, it would amount to little more than a scholarly treatise on the subject for people who neither chop nor stack nor burn wood themselves.

Wood isn't something much thought about or talked about in Norwegian public life, at least not until the larger connections are made to the goal of a society based on bioenergy. Yet wood will always resonate at some deep level inside me and my compatriots, because our relationship to fire is so ancient, so palpable, and so universal.

That's why this book is dedicated to you, Ottar. You remembered something the rest of us keep forgetting: that winter comes around each year.

Lars Mytting
Elverum, –24°F (–31°C)

CHAPTER 1

THE COLD

In need of fire he is, he who steps inside,
numb with cold knees.

—*Håvamål (Sayings of the High One)*, from the *Poetic Edda*, orally transmitted Old Norse mythological poems recorded in the thirteenth century

It was the difference between being frozen and being warm. The difference between ore and iron, between raw meat and steak. In winter it was the difference between life and death. That is what wood meant to the first Norwegians. Gathering fuel was one of the most crucial of all tasks, and the calculation was simplicity itself: a little, and you would freeze. Too little, and you would die.

Perhaps in the course of a few thousand years of frost and suffering a uniquely Nordic woodburning gene has evolved, one that is lacking in people living in more clement parts of the world. Because wood is the reason the northern peoples are *here at all*—without it these cold regions would simply have been uninhabitable. A mere century or so of fan heaters has not been enough to wipe out that debt of gratitude—and the joy that harvesting firewood brings may well reflect the awakening and activation of that gene, something that connects us through the ages to the gatherers we are all descended from.

For thousands of years wood was *serious business* in Norway. From the earliest times, people in the north have chopped green wood and dried it in preparation for the coming winter. Wood has left its mark on the Scandinavian languages. In Swedish and Norwegian, the word for "firewood" is *ved*, and the

Old Norse word for "forest" is the almost identical *viðr.* Trees meant heat, and since time immemorial people have gathered around open fires in their camps, and later around fire pits, with the smoke escaping through a vent in the roof or tent.

Of course, in former times wood was crucial to the survival of people all over the world, for heating and preparing food. It is our most ancient source of energy, its uses and traditions subject in the main to two conditions: what kind of forests there were, and how cold the winters were. In the years around 1850, for example, the one million inhabitants of Paris consumed annually some three hundred thousand cords of wood. If in our times Scandinavia has become an especially interesting region in which to study the history and culture of heating with wood—bearing in mind that the use of firewood here has *increased* hugely over the last thirty years—the principal reasons are these: We have a wealth of forestland; our tradition of woodburning has never been broken by the adoption of coal burning or any other means of obtaining heat; Scandinavian countries have been in the forefront of the development of clean-burning stoves with minimal pollution; and, perhaps the single most important factor of all, we cannot modernize our weather. Up here in the north it is still cold.

The Pleasures of Chopping Wood

Wood is chopped, dried, and stacked in fairly similar ways across the Scandinavian Peninsula. Consumption in Norway, Sweden, and Finland is on average 660, 750, and 860 pounds (300, 340, and 390 kilograms) per capita, respectively. Populous Sweden alone goes through three million metric tons of wood a year. Even in oil-rich Norway, an astonishing 25 percent of the energy used to heat private homes comes from wood, and half of that is wood chopped by private individuals.

So the consumption of wood in present-day Scandinavia is not great.

It is *enormous*.

How big? Well, if we take as an example the annual consumption in

PREVIOUS LEFT Birch has always been regarded as the queen of firewood in Norway. It grows tall with few branches, and splits easily. This is a meticulously cared-for forest of birch near Fåvang in Gudbrandsdalen. Most of the trees were planted twenty years ago, and undergrowth has been cut away at regular intervals.

PREVIOUS RIGHT A drying bin made of iron mesh is a good supplement to woodpiles—it is perfect for twisted wood that is difficult to stack.

OPPOSITE Mountain birch in a fine, robust square woodpile stacked by Eimund Åsvang of Drevsjø.

Norway, which is 1.5 million metric tons, assume that each log is twelve inches long, and pile all the logs in a stack 6.5 feet high (ignoring the considerable risk of collapse), we would have a woodpile 4,474 miles long, stretching all the way from Oslo to the center of the Democratic Republic of the Congo. It might be simpler to stack the logs on a flat surface. If the pile was still 6.5 feet high, it would cover an area of about two square kilometers.

There's no mistake. The calculations come from the number crunchers at the Central Bureau of Statistics of Norway, who often receive astonished feedback at the sheer amount of wood Norwegians go through. In fact, the average annual consumption of a modern Norwegian is 20 percent greater than that of the Parisian of 1850. It might be easier to grasp if we imagine that to transport the timber that makes up a load of 1.5 million metric tons, we would need about two thousand freight trains, with each train hauling twelve cars. It still sounds like an incredible amount, but one-third of our country is covered in forest and, if we take a bird's-eye view of things, that pile of wood stretching to Africa is a mere drop in the bucket. In fact our annual consumption of wood is just 12 percent of the mean annual growth, and less than 0.5 percent of the volume of standing trees in Norway.

Here is a good place to expand our horizons a little and note that we Scandinavians, with our skis and our thick winter jackets, are not in fact the world record–holders in the consumption of wood, nor does that prize go to the fur-clad Russians of Siberia; it belongs, rather, to the inhabitants of tiny Bhutan, whose average annual consumption is a staggering 2,000 pounds (about 900 kilograms) per capita. Ninety percent of all the energy used in heating and cooking comes from wood, and in the villages of the Bhutanese countryside the consumption is 2,750 pounds (about 1,250 kilograms) per inhabitant. The Bhutanese chop down the equivalent of the annual growth, so consumption at this level is both an environmental and a social problem, since the country teeters constantly on the verge of a wood shortage.

In former times, large parts of Europe also experienced crises arising from a similarly fraught state of affairs. A few centuries ago the amount of wood used in smithies, in building houses, and in shipbuilding was so great that vast expanses of the continent were completely deforested, and shortage of wood became a chronic problem in many lands. Even Sweden suffered. Dwelling places in those days were heated by open fireplaces that had to be kept going night and day, imposing on the occupants what we might nowadays call an open-plan interior, in which all members of the household had access to the central fire.

Open fireplaces do not give off much heat and require enormous quantities of wood. In 1550 it took more than thirty-three thousand loads of wood to keep the Swedish king John III and his courtiers warm during a winter spent at the castle in Vadstena. Large areas of Swedish woodland were almost clear-cut to provide fuel for the great ironworks, and by the eighteenth century the supply of wood had almost dried up. But the Swedes are a resourceful people, and the government commissioned two talented engineers to design and build a more efficient stove. In the space of a few months they had constructed the Swedish *kakelugn*, the famous tiled stove still in widespread use today. On the first working drawings, from 1767, it was specifically noted that the stoves had been designed "to economize on wood."

In Norway it was only the oak forests that were clear-cut, and the population was never large enough to threaten a shortage of wood. Cast-iron stoves were the norm in both Norway and Denmark—the oldest surviving one is from 1632. Access to forests was good in Finland, and coal never became a major source of heating energy there either. Not until the coming of electricity and heating oil in the twentieth century did the consumption of wood in northern Europe start to drop, especially in the towns and cities. In Great Britain the adoption of coal was made necessary by the disappearance of the great forests. Oscar Wilde, a firewood enthusiast, observes in chapter 3 of *The Picture of Dorian Gray* that the advantage to an English gentleman of owning a coal mine was that the income enabled him "to afford the decency of burning wood on his own hearth."

Security in a Time of Crisis

Then came the Second World War—and wood demonstrated to the full its extraordinary value in a time of crisis. In occupied Norway, the availability of coal, coke, and heating oil sank dramatically. Sales of wood in Norway in 1943 were four times what they had been in 1938, and there was a huge increase in the number of trees felled on the farms. Energy supplies in Finland were still largely based on woodburning, and during the war years the Finns created huge stocks of firewood. More than ten million metric tons were felled each year, and Hakaniemi Square, in the Finnish capital, Helsinki, was filled with wood every winter. The square was one mile long, and the stacks thirteen to sixteen feet high, so these may well have been the largest woodpiles ever built.

The war years were a timely reminder of the value of renewable energy that is locally sourced, and in the postwar years the woodburning stove played a decisive role in the rebuilding of northern Norway. In 1946 the government urged the major woodstove manufacturers to forget the more lucrative foreign markets and devote their efforts to making stoves for use up in Finnmark, Norway's northernmost county, up above the Arctic Circle. The logic was simple: If there were no stoves, then no houses would be built, and people would be unwilling to move back to revitalize the devastated region.

Some of the largest woodpiles in the world were those built in Hakaniemi Square, in Helsinki, during the Second World War. The wood was stacked every year in piles thirteen to sixteen feet high. Finnish soldiers developed an attacking strategy that they called the "motti maneuver," after *motti*, the Finnish word for "a cord of wood." This photograph was taken sometime between 1941 and 1944, years in which the annual timber harvest in the country was close to a record twenty-five million cubic meters.

The return of peace also saw the return of manufactured and convenient forms of heating. Contemporary advertisements for the electric radiator are eloquent on the spirit of the times. Along with the dishwasher, the vacuum cleaner, and the linoleum floor, the radiator was a part of the modern era of convenience. Finally the family was liberated from the danger of fire, splinters in the fingers, soot clogging up the air bricks, emptying the ash pan, the constant attention required to keep the fire going, the misery of living in a cold house with a stove that went out during the night, the endless trips out to the woodshed in shirtsleeves to fetch more wood, the stream of injunctions from the chimney sweep on the subject of tiny cracks in the chimney and broken firebricks. People did not have to rise in the middle of the night to put more electricity in the radiator. It must have felt wonderfully modern to be half roused from sleep by the tiny click of the thermostat, distantly recall the bad old days when you might have had to get up and go for more wood, and now simply turn over on your pillow and go back to sleep.

It isn't surprising that the use of firewood fell into decline during this era. The chain saw didn't become widely available until twenty or thirty years after the war, and firewood to a very large extent meant physical labor. Nor were woodstoves as efficient as they are today. The cheaper radiators required less maintenance and could retain their warmth through the night. Houses were often poorly insulated and needed some simple source of background heat throughout the day, a task for which electricity was ideally suited. The sale of firewood sank dramatically throughout the postwar years. In the 1970s electricity and oil were so cheap that only those with access to free wood used it as a source of heat. Heating with firewood in the northern regions reached an all-time low during this period—and then it began to rise again, and it has kept on rising. Today, consumption in both Norway and Denmark is *ten times* what it was in 1976.

Several factors combined to encourage the return of the wood fire. Oil and electricity prices rose, clean-burning stoves made their appearance, and manufacturers started putting an emphasis on design. Shortly afterward, fears about climate change and an unstable world economy joined the list. Woodburning, with its half-forgotten virtues, emerged once again in all its simple glory. Being a carbon-neutral renewable, wood was given a clean bill of health and embraced by the environmentalists.

Another factor, emerging as if part of some completely different plan, was the way technology suddenly appeared to favor the woodcutter. All farmers had tractors now, and other people had cars and trailers. Good chain saws became available at an affordable price, and the wood processor made its appearance, its technology revolutionized and on sale at a price that put it within reach of every farmer, large or small. This sturdy contrivance, usually run from a tractor, cuts, splits, and transports wood along a small conveyor belt into sacks or onto a pallet. The wood processor makes it possible for one man alone to handle large logs and cut a generous amount of firewood quickly, and profitably. Farmers everywhere began to see the potential for income in the sale of firewood and banded together as the Association of Norwegian Firewood Producers, an organization that now has more than forty-five hundred members. They endorsed a series of quality criteria developed by the Norwegian Wood Standards Board (the board, a pioneering enterprise, is now the reference point for a number of other European countries), and the result is that access to good wood at reasonable prices is better than it has ever been.

Yet the return of the log fire can hardly be reduced to a matter of money. Many people feel that a living fire gives a rich experience. We are drawn to the fire, just as we once gathered around the flames in former times. For many there is a qualitative difference in the heat supplied by a radiator and that provided by a woodburning stove. A stove can glow with heat. Your feet won't get warm when you turn on the inverter, and a radiator has to be on for quite some time before it will drive the chill from a cold house. Electric radiators seldom deliver more than two thousand watts, whereas even a small woodstove is easily able to generate six thousand watts, and many stoves as much as fourteen thousand watts. Scientifically speaking, there is no measurable difference between the heat generated by electricity and that produced by combustion, but the body reacts in a different way to the more intense heat from the stove, not least because modern fireplaces with glass doors radiate heat. An ordinary electric radiator or heat pump warms only the air in the room, but flames and glowing embers release electromagnetic, infrared radiation that has much the same characteristics as sunlight. Warming occurs *in* the skin and the body as the radiation arrives, with an immediacy and an intensity that bring a feeling of well-being and security. The indoor climate is also slightly changed. The consumption of oxygen encourages a degree of air circulation, and the stove absorbs a quantity of dust. These factors, combined with the smell of wood and a little woodsmoke, and the sight of the ever-changing play of flames, connect us with the primordial magic of the fireplace.

Something else to consider is the way the woodburning stove brings people into a very direct relationship with the weather. You are your own thermostat, you are the connecting link between the subzero temperatures outside and the relative warmth within. When you heat with wood you have to go out to the woodpile, come back in again, and start your fight against the cold. It's bitter, and it bites, but you can do something about it. In this one small but vital arena you are in touch with the bare necessities of life, and in that moment you know the same deep sense of satisfaction that the cave dweller knew.

It may also be the case that we have become modern enough to look back and appreciate things that the generation before us did not value. Things go around. With the advent of kitchenware made out of hard plastic, the wooden bowls and utensils were consigned to the flames (those that survived are now sold everywhere as "rustic antiques"), and people were happy to see them go up in smoke. At last they were rid of that heavy, crude rubbish that was impossible to clean properly. They probably felt pretty much as we do today when we finally get rid of a sluggish old computer. The generation before ours covered the oak floors with linoleum and hid the intricate beauties of the Swiss chalet–style houses behind sheets of plywood. For us, their time has come again.

Yet heating with wood is not a nostalgic gesture on the part of the Scandinavians. It is both a source of energy and a deeply rooted part of our culture. The way a person cuts and stacks wood can tell you a great deal about him, and the woodpiles you see round about in the countryside are a reminder that wood is the connection between the forest and the home. It's a modern and practical form of the old Romantic nationalism, as much a part of the soul of our countries as cross-country skiing, mediocre local newspapers, and elk hunting.

But this alone is not enough to explain why so many swear by a Stone Age method of heating a house that can boast a fiber-optic broadband connection. The main reason for the increased use of firewood is an entirely pragmatic one: Woodburning has been modernized and integrated with other sources of energy. Wood has a particular role to play as a sort of national insurance against the cold. One of electricity's shortcomings is that technical problems can cause it to fail completely. Large parts of Norway experience long periods of extreme cold in the winter, with temperatures that hardly warrant a paragraph in the local newspaper until they reach minus forty degrees (a reading that is the same whether your thermometer is Celsius or Fahrenheit). In circumstances such as these, power cuts of even a few hours' duration will precipitate a local crisis. Many communities, especially coastal ones, draw their power from a single electrical system. If this fails, then there is no better or more universal remedy at hand than a good supply of firewood to provide the heating, boil the water, and cook the meals. In January 2007, when the remote community of Steigen, in the county of Nordland, found itself without power for six days during a period of especially cold and stormy weather, it was the woodburning stove that enabled residents to ride out the storm in relative comfort.

For this reason every house in Norway exceeding a certain size is obliged by law to have an alternative source of heating, which in practice means a woodstove. The requirement comes not, as one might think, from the Building Standards Department, but from the Directorate for Civil Protection and Emergency Planning, and the explanation for this is quite simple: A woodstove and a supply of firewood will prevent the spread of panic and the possibility of having to evacuate homes. And this is not merely because wood is a source of energy; it is because wood is an extremely *adaptable* form of energy. It can be shared with your neighbor, it doesn't leak, it doesn't need cable, a match will light it, it can be stored for year after year, and even inferior-quality wood will still do the job for you. There is peculiar security in the fact that this is energy in solid and tactile form. You can carry it into your house and know that the weight of what you are carrying represents exactly the amount of heat you will be getting.

Henry David Thoreau is frequently quoted on the subject of our relationship to wood. In 1845 Thoreau went to live in the forest because life in modern American society had become too hectic for him (that's right, in 1845). In *Walden* he wrote: "It is remarkable what a value is still put upon wood even in this age and in this new country, a value more permanent and universal than that of gold. After all our discoveries and inventions no man will go by a pile of wood. It is as precious to us as it was to our Saxon and Norman ancestors."

It was Thoreau too, again in *Walden*, who pointed out that wood warms twice over, once when you chop it, and again when you burn it. He might have added the warming effects of splitting the logs, and of stacking and carrying them, but that would not have harmonized with his philosophy.

Woodburning is part of the cultural heritage of every Norwegian, but that does not make us model exponents of the use of environmentally aware bioenergy. Like city dwellers everywhere, our urban population has accustomed itself to the idea that everything is available at all times, in all places, and at the press of a button. Whenever a cold snap comes along, firewood merchants see firsthand the desperation that comes when electrical heating isn't enough for the job and the wood has run out. Decent, law-abiding citizens can turn into bullying line jumpers, and honest men and women into deceivers and liars, as they try to wheedle their way to a few more sacks of the precious stuff. (Many of Oslo's wood merchants make it a practice to give priority to the elderly in a time of crisis.) As soon as the cold settles, the first two people interviewed on the radio are inevitably the press officer of some hydroelectric-power company, who informs listeners that water levels in the reservoirs are at a record low, and the wood merchant, who reports, "People have forgotten to stockpile. They don't buy until it actually turns cold."

Pollution

One of the great issues of the age has to be addressed here: Is it possible to burn wood and still be a good environmentalist? Woodburning stoves release a lot of carbon dioxide, as much as four pounds (2 kilograms) for every two pounds (1 kilogram) of normally dry wood, and yet heating with wood is accepted as a source of green energy by almost all experts, for this simple reason: Trees absorb carbon dioxide as they grow, but sooner or later release the gas again. A tree burned in a stove releases the same amount of carbon dioxide as it would had it been allowed to die and rot.

OPPOSITE "Near the end of March, 1845, I borrowed an axe and went down to the woods by Walden Pond, nearest to where I intended to build my house. . . . The owner of the axe, as he released his hold on it, said that it was the apple of his eye; but I returned it sharper than I received it."—Henry David Thoreau, *Walden*, 1854. So strong was Thoreau's longing for a spartan life free from the obligations of possession that he even *borrowed* his ax. The ax shown is a Swedish Gränsfors American felling ax. It is not unlikely that Thoreau's ax had much the same proportions.

GRÄNSFORS BRUKS
SWEDEN

The poet Robert Frost touched on the phenomenon in a poem that describes an abandoned, rotting woodpile, which warms with "the slow smokeless burning of decay."

Woodburning is an efficient way of making use of one of the world's best sun catchers, simply by allowing one of nature's processes to take place indoors. Forests are able to absorb an extraordinary amount of carbon dioxide. The problem is that trees don't live forever. Sooner or later—in the case of certain kinds of trees, within thirty years; in other cases, not for centuries—the tree dies and starts to decay. The same gases are then released in the same quantities as they would be by burning. The only way to prevent this emission would be to fell the trees and keep them in perpetual storage in some vast woodshed, or to use them as building material. But wooden houses also have a finite life span, after which they either rot or are demolished. The general consensus among Norwegian scientists is that most of the carbon dioxide is absorbed while the trees are young and growing, so forest renewal can raise the absorption of carbon dioxide for a time. This is true for a young forest environment where regrowth is good; there are of course a number of exceptions—for example, the fact that carbon dioxide is also stored in the forest floor.

Thus, heating with wood does not lead to an increase in greenhouse gases, provided that consumption and renewal are balanced. The question, then, might seem to be merely whether or not to accept combustion as a source of energy. Unfortunately, it's not that simple. Woodburning's single great problem, especially in urban areas, is the pollution that comes out of the chimney. A woodstove is not like a radiator that can be turned on and off at the flick of a switch—a certain amount of theoretical insight is necessary to make sure combustion takes place cleanly and correctly, and burning in the wrong way can be a source of serious local pollution.

In Norway in 1982, a report appeared that shocked the country. A study of atmospheric pollution had been carried out in Elverum, in Hedmark County, an area dense with pine forest and where temperatures can sink to minus twenty-two degrees Fahrenheit (minus thirty degrees Celsius) and stay there for weeks on end. Elverum, a small town, has a record number of woodburning stoves, and the results of the study showed that heating with firewood generated as much atmospheric dust as the traffic passing through central Oslo—and this in the days of those two major producers of atmospheric pollution, the studded winter tire and leaded fuel.

The source of the problem soon became apparent: The pollution was the worst during periods of milder weather, because people were in the habit of loading the stove with wood for the night and then damping down the air supply so that the logs would burn slowly through the night and give off a steady background heat. Some would add a bit of green wood to slow the process down still further.

For Norwegians, this was a wake-up call and the beginning of a search for a new design of stove that would greatly reduce the rate of harmful emissions. State-funded research institutes joined forces with stove manufacturers to design stoves that would burn more efficiently and reduce the amount of atmospheric dust produced. As long ago as the 1960s the Norwegian firm Jøtul, which exports stoves all over the world, had made prototypes of what are now known as "clean-burning" woodstoves; with state backing, progress was swift. A law was introduced in Norway in 1998 stipulating that all newly installed stoves be clean burning, and Norway's emission regulations remain among the most stringent in the world today. A number of campaigns also encourage people to learn the correct use of woodstoves.

Research figures show that woodburning in Norway today is a dramatically cleaner business than it was thirty years ago. The emissions from good-quality wood burned in a modern, clean-burning stove using the proper techniques amount to less than 5 percent of the emissions from the old stoves when these were used heedlessly; typically they are in the region of 4 to 5 grams per kilogram consumed. The very best Norwegian and Danish stoves give off a mere 1.25 grams of atmospheric dust per kilogram (2.2 pounds) of wood, compared to 40 to 50 grams or more in the old ones. They combust more efficiently too—in some models as much as 92 percent of the energy potential in the wood is used. In the United States, the Clean Air Act requires that woodstoves be certified by the Environmental Protection Agency, and the government instituted new regulations in 2015 regarding emissions standards. The website epa.gov/burnwise offers guidance on the purchase, installation, and maintenance of the various approved woodburning appliances.

And yet woodburning remains responsible for more than half the atmospheric dust released in Norway, leading to successive campaigns aimed at reminding people to use the stoves correctly. In several districts, local governments offer financial incentives to people to convert to clean-burning stoves.

Norway is not alone here. Much of the research into the harmful effects of using woodstoves in an urban environment comes from Christchurch, New Zealand's oldest city. In the older parts of Christchurch, firewood is still an important source of heating. The geography of the town means that it is vulnerable to smog, and much of the atmospheric pollution comes from the use of wood as a fuel—observable in the form of a marked increase in cases of respiratory illnesses. Clean-burning stoves are obligatory, and wood merchants face a fine if they sell wood with a moisture content above 25 percent.

Of course, in a modern society one doesn't expect users to convert en masse from electricity to wood. But wood can be a fantastic *component* of any energy plan. The problem facing a society that relies solely on electrical power is that that capacity has to be as great as the maximum consumption during the coldest periods. In Norway, woodburning stoves play an important role in relieving the demand for electricity when the real cold comes along, and it is an acknowledged fact that wood reduces the need to expand the national power grid. The simple logic governing electricity's use is entirely in the hands of the householder: In periods of cold the price of electricity rises, and electric radiators may not be up to the job anyway.

Even when compared on an equal footing with other sources of heat, wood comes out remarkably well. Electricity from hydroelectric power is usually regarded as the cleanest source of energy, but establishing it involves a major assault on the landscape in the form of dams, heavy machinery, and power lines. And a power cable often loses 20–30 percent of its efficiency on its way from the turbine to the electrical socket. Coal, fuel oil, and kerosene are not renewables, and their production requires a huge industrial superstructure, with a lot of the energy going into both production and transportation. Electrical current passes through overhead wires, and if these suffer damage, the energy is gone. If a fuel truck overturns or an oil tanker springs a leak, environmental damage of a different kind is done. The same goes for pellets—they're renewable, but they have to be made in a factory. Wind turbines are not always a popular feature of the landscape, and nuclear power, though efficient, can lead to disaster.

By comparison, the infrastructure required by woodburning is almost touchingly simple. A cast-iron woodstove will last for at least forty or fifty years, and in most cases it will be fed with locally sourced wood. But more important than that is how simple wood is to obtain, how little bureaucracy is involved, and how little energy is expended: A tree is felled, chopped, and split, and then the logs are stacked to dry. A new tree starts to grow where the old one stood. After a few months' drying you've got 4.2 kilowatts of energy for every two pounds (one kilogram) of wood. (Coal may give almost double, but you need to factor in the costs of extraction and transportation.)

For personal use, all that is really needed are a chain saw, an ax, and a car with a trailer. Two reasonably fit people can create a stock of wood worth twelve thousand kilowatts with about one week's work. This will probably take about seven or eight trailer loads, and if the wood is not too far away, twenty to thirty liters of fuel, maybe less, will cover the transportation and the chain saw.

The whole thing is quite straightforward. No paperwork is involved, and it is within the capabilities of young and old alike. The work is repetitive but never boring. It will bring as much pleasure to the company director who wants to unwind and let his thoughts wander as to someone in search of a straightforward job to do. In Norway there are many woodyards run by local governments where people with learning difficulties and other disabilities are able to contribute as best they can, an arrangement that also fosters the use of locally sourced bioenergy.

When it comes to woodburning on the industrial scale, involving harvesting machines and the large-scale extraction of biomass, the picture is not so simple. The time difference between absorption and emission has to be factored in: If you fell and burn very large amounts of old forest, the amount of carbon dioxide released will not be taken up again until a corresponding amount of new forest has grown. Where the disparity is considerable, the climate may be adversely affected during the time it takes for the balance to be restored. Additional considerations include emissions from the machinery itself and the transportation that is part of large-scale industrial felling.

The disadvantage of wood for the consumer is that it requires effort. It is heavy and has to be carried to the stove, which in turn has to be regularly attended to and which will go out after a few hours. This problem is much reduced by taking the simple step of purchasing one of the many stoves that store heat. One of the most interesting modern developments are boilers able to heat buildings up to the size of large office blocks by means of waterborne heat. Many of these are fed and adjusted automatically, and the most recent designs pollute very little.

A minor point worth noting is that local green energy is not a contentious issue on the larger political stage. Countries that depend wholly on oil, coal, and other forms of fossil fuel guard their resources carefully. But no one has ever gone to war over a firewood forest, and no species of seabird has ever been drenched in oil because a trailer load of firewood ended up in a ditch. A woodpile might not stop a war from breaking out, but simple, local sources of energy are not the stuff of violent conflict.

Regardless of the scale—whether we look at woodburning as culture, as bioenergy, or as a way of getting closer to nature—the world's oldest source of energy still has a great deal to offer us. Crusaders for the cause of "intelligent woodburning" in Norway swear by the truth of their motto: At last, something we need that actually does grow on trees.

NORDSKOGBYGDA: THE BACHELOR'S FARM AND THE MODEL FOREST

The first snow has fallen, and a set of car tracks through the forest shows the way to Arne Fjeld's home in Nordskogbygda, in southern Norway. Arne is a small farmer with an interest in local culture and history, and a frequent lecturer at the Norwegian Museum of Forestry, in Elverum.

The woodburning season is well under way now, and stoves ancient and modern provide a fine background heat in the farmhouse. It was built in 1870, and the sparklingly clean original windows have the sort of small irregularities that only old window glass has. The hunting terrain starts right outside the front door, and today the shotgun is leaning in the porch. Sunlight flashes on three rifle cartridges lying on the windowsill.

"I think," says Arne, "that the more of a city person you are, and the easier it is for you to get what you want from a shop that's open around the clock, the more astonished you are by these enormous woodpiles that you see out here in the countryside. It's probably more or less incomprehensible for people from the warmer parts of the world too. In Australia the Aborigines gather dry branches whenever they need a fire for cooking food. But here in Norway we might as well admit that, when it comes to the climate, we live in a developing country. In the old days the long, cold winters forced us to plan well ahead. There was no choice. You had to stockpile enough wood to last you through the whole winter, and add a little extra to cover any unforeseen circumstances. That's just the way things were. You had to have enough, otherwise you and your family would freeze, and if worse came to worst you'd all freeze to death. I think all this has left its mark on Norwegians. Until the day came when society was organized in such a way that the strong looked out for the weak, and the smart tidied up after people who made a mess of things, or had a bit of bad luck—well, you had to look out for yourself, you had to make sure you were well provided for. And if winter turned out to be three weeks longer than usual, well then, you'd freeze."

Arne gets his wood from several parcels of forest close to the farm, but one in particular holds a special place in his heart. He's lavished a lot of care and attention on it over the years, and it also has a rather special history. This piece of land was originally tied to a small farm nearby. A woman lived there with her son, born to her when she was very young. The boy never married, and the two of them ended up living there together for the rest of their lives. This little household was more or less self-sufficient from the yield from its single, well-cultivated field.

Arne Fjeld's model wood was grown in an abandoned field. Over the years these slender trees with their smooth bark have thrived under his watchful eye.

The mother lived to a ripe old age, by which time the son was so old that he moved straight into an old folks' home. Arne expressed an interest in buying the place, but years passed without anything happening, and the farm was allowed to run down. Finally, after twenty years, Arne's bid was accepted and he bought the place. By that time the field was overgrown with birch, helped by the fact that the previous owner had plowed the field just before moving out. The birch had taken root all over the field, and Arne had to abandon his original plan to sow wheat there.

"But then it struck me," says Arne, "that this was a unique opportunity to create the perfect woodlot. The trees were all the same age and the soil was of outstanding quality."

And so Arne started to cultivate the wood with as much loving care as though each tree were a rosebush. He thinned it out three times until he had trees of varying heights. It was all hard work, and his wife would smile and shake her head in disbelief when he got up at the crack of dawn to shake the snow off the trees so they wouldn't break under the weight. But Arne and his wife have every reason to be proud of the results. The birches stand as straight as flagpoles. Many have thirty feet of trunk completely free of branches. Arne takes birch bark from his best trees in addition to using them for firewood. The bark is as smooth as leather and fetches a good price.

The main farmhouse dates from a time when the woodburning stove was the only heating option. Each floor of the house has its own magnificent tiered

Arne gets all his wood from the forests surrounding his farm and makes sure his stacks are well aired, with ample space between them, to ensure proper drying.

stove, and there is a huge chimney with a ladder mounted on the inside large enough for a sweep to climb down it and emerge again through the sweeper's hatch, about one square meter in size.

"I don't think people in the old days had a particularly personal or romantic attitude toward wood," says Arne. "Most farmwork in those days was heavy, backbreaking stuff, real drudgery. Wood was just something that had to be dealt with, along with a lot of other jobs waiting to be done. A huge amount of wood was needed to meet the needs of a whole farm. The geography of this farm means that the snow lingers for a long time, so before farmwork was mechanized, spring was an extremely busy time of year."

Work was nonstop through the summer and there was no time to gather the wood until well into the autumn. By that time a man would have been working almost the whole year chopping and piling it, but it wouldn't be transported to the farm and chopped into logs until the autumn. It meant that quite often the wood wasn't dry in time for winter.

"It used to annoy me so much when I was young," says Arne. "Probably

the reason I'm so passionate about wood now is that my father never made a priority of it. I remember the wood was often frozen, you had to break the logs apart and thaw them out under the stove. The kitchen floor sloped and we had little rivers of ice-cold water with sawdust in it running down the floor. Now, at my age, if I hear a log hissing, it irritates me. So I do the felling and chopping in the spring. And I absolutely love doing it. I use all different kinds of wood and notice the different ways they burn in the stove. I examine each log, and if it has a nice grain I put it to one side and use it to make a knife handle. That's the way I am now. I can't bear to let anything go to waste."

Splitting logs is the part of the job Arne enjoys most.

"It's therapeutic. And it's not a very complex job. Routine, really, but not boring. So many things that happen in our everyday lives bother us and cloud our day. Often, if I've been to a meeting and gotten worked up about something or other, I might go around thinking of all the things I should have said. But then, when I'm standing by the chopping block, I don't think about any of it anymore. My mind is never so pleasantly empty as it is when I'm chopping wood."

CHAPTER 2

THE FOREST

Ah, they were good days, chopping cordwood with Inger standing there watching; they were the best days.

—Knut Hamsun, *Growth of the Soil*, 1917

Like many another man, Isak, the main character in Knut Hamsun's novel *Growth of the Soil*, was at his happiest when in the forest. And just to preempt any accusations of male chauvinism: Isak's wife, Inger, didn't stand there watching him because she was lazy, or awestruck at the sight of a man wielding an ax. Hamsun described it that way because Isak derived no pleasure from doing work that no one appreciated or was able to see.

So pity the man who chops wood for himself alone.

There is nothing quite like it: heading off into the forest with the snow still lying in patches around the trees, and the spring air thin and cold. You shrug off your knapsack, fuel your trusty chain saw, and away you go. Easy does it at first: Start with the smaller trees to make it easier to get to the big ones. Get ready for the first joy of the year: one of those tall birches close enough to its neighbors to compel just the right degree of struggle in stretching toward the light, so that it stands there now tall and slender and white. Pull-start your chain saw, set it to the trunk of the tree, and before long hear the great *swish* as the tree falls in more or less the intended direction. Turn off the chugging two-stroke engine, lift the visor of your helmet, and let the great silence embrace you as you stand in contemplation of a fallen tree that will, come winter, keep you in warmth and light.

Nature is bountiful in so many ways, but the work of the woodcutter is always rewarded. For him, unlike the fisherman, the hunter, or the berry picker, the fruits of his labor are guaranteed. Cutting wood is proper work, something people can do together, and there can be few jobs where the results are visible so soon after the effort has been made.

Each tree is worth its weight in kilowatt-hours. Even the humblest stick can be used. It's like picking up small change: One day all those coins will add up to a note. It's a good feeling. You're not wasting anything. The trunk of some puny dwarf spruce will make welcome kindling on a day when December gives way to January, the outside thermometer shows minus fifteen degrees Fahrenheit (minus twenty-six degrees Celsius), and the dishwashing liquid clings to the inside of the bottle.

Where to Work?

Not many people have a timber forest of their own, but there is plenty of communal woodland across Norway accessible to the individual woodcutter, and many local authorities allow felling at reasonable rates. Nowadays most forestland owners use logging machines, and over the last thirty years the number of those employed in the forestry business has fallen to half of what it once was. This development has left good openings for a woodcutter with a strong back, who is encouraged to thin the woods, or to cut in terrain that is either too difficult or too sparsely treed for the machines. A well-thinned forest helps the commercially valuable timber grow taller and straighter, so the contribution of the individual woodcutter is welcomed by the owner.

Another option is to get permission to work along roadsides, close to power lines, the borders of fields, old pastureland, and along creek beds. Many forest owners will also allow private individuals to clear up after the logging machines have clear-cut an entire area. A lot of timber gets left behind in the process—tops and twisted or broken trunks. There isn't the same pleasure as working in a quiet timber forest, but there is less effort involved. Remember to

PAGE 36 Hedmark is the largest pine district in Norway, and these young, slender trees will grow into excellent building materials, perfect for log houses. Younger aspen and birch growing between the old trees will be cut for firewood.

PREVIOUS LEFT Classic elements in a Norwegian private timber harvest: retaining straps, a battered trailer, and a farm track in the early spring.

PREVIOUS RIGHT Winter is the perfect time for forest work. Here, spruce has been cut to logs immediately after felling. It will be split on-site and moved out of the forest on a sled.

leave the dead spruce and pine where they are—these are important elements of the habitat for insects and help preserve the balance of species.

In thick woodland, most types of trees will grow taller and develop fewer branches than trees that stand relatively alone and get a lot of sunlight (these tend to be on the short side and to develop branches all the way up the trunk), for where space is at a premium trees have to compete with their neighbors as they stretch toward the light. These features are particularly noticeable in trees that grow on steep slopes or in depressions. Tall, straight trees such as these make ideal firewood. They are easy to fell, simple to extract, split easily, and make sound structural elements in building a woodpile. Trimming twisted and thickset evergreens, on the other hand, is hard and time-consuming work.

The woodcutter is not interested in the biggest trees. Most kinds of trees, by the time they are between twenty and forty years old, have grown to a size that makes them usable but still manageable. Trees older than that tend to be so big and heavy that it takes a lot of hard work to drag them from the felling site to the trailer.

Into the Forest

A degree of self-control is essential on your first few trips into the timber forest, and a "rule of thumb" means something quite different to a chain-saw owner than it does to an office worker. Lumberjacks in the old days suffered appalling injuries, and their modern-day successors with chain saws who approach the task with anything other than extreme caution risk similar serious injury—the loss of a thumb at least. At full speed the chain can run at forty to fifty miles an hour, and with a standard bar size the user is in fairly close and constant contact with some eight hundred whirring and rapacious metal teeth. A whole book and a certain amount of practical instruction are needed to teach the art of tree felling; in Norway it is generally accepted that the way to learn the proper use of a chain saw is either in a course or from an experienced user, not least because the teaching will take account of local conditions and the types of trees the user is likely to encounter. Such courses in the safe use of chain saws are offered in several parts of the country, and chain-saw manufacturers provide detailed instructional booklets and short, expert videos (also available on the Internet) that show the right way to go about it.

Among the basics that are stressed from the very beginning: Seek out an experienced user, never take dogs or children with you, start on small trees, and take extra care if a wind gets up. Note how even a small gust of wind can make a large tree sway. It is difficult enough at the outset to get a medium-size tree to fall in the right direction, and once a wind gets up, a remorseless element of

A good workout and useful too: Jo Gunnar Ellevold uses his chain saw to strip-bark newly felled birch in a training corral at the Terningmoen army camp. The exercise provides winter fuel and prevents overgrowth.

uncertainty enters the process. A principle always observed is that the tree is felled in *two* cuts. The first is the sink, or directional, cut, which penetrates about a quarter of the way into the trunk. The tree will later fall over this cut. Then, from the other side, the felling cut is made; it is sawed in the direction of the sink cut, but stops about an inch before the two cuts meet. The remaining section functions as a hinge, ensuring that the tree falls to the ground in a controlled fashion. The direction and shape of the felling cut can be adapted to ensure that misshapen trees fall in the desired direction; attempting to learn this particular art in theory can be likened to trying to learn the tango by correspondence course. And even if the sink cut is made perfectly and everything else has been done correctly, nothing can stop a tall tree if a gust of wind moves it in the wrong direction. In the words of the poet and wood enthusiast Hans Børli (who, in addition to writing verse, worked as a logger his entire life): "This superficially simple and straightforward activity is in fact full of subtleties, little tricks, and arts that can be learned only in the hard school of experience. . . . You've got to be able to get that tree down on a dime, otherwise you're stuck and you've got a hell of a job afterward with the peavey."

So novice woodcutters start out cautiously and learn from experience as they go along. They learn how to use the tree's own natural inclination to work out where it wants to fall based on its center of gravity and the slope of the land. The danger is that if one is not very careful the tree will get caught up among the other trees, the chain saw might get stuck—and any number of other difficulties and dangers can arise. A lot of injuries occur while limbing branches—again, an argument in favor of taking a course or at least watching the manufacturers' instruction videos. Wood won't warm much when bits of your body are lying in a container outside the emergency room of your local hospital.

Woodcutting is demanding work. Norway, a country that in the 1960s had its own advisory body for the burning of wood chips, naturally also has a body for studying the relationship between nutrition and physical activity. Its findings show that working in the forest is one of the most strenuous of all forms of activity. Lugging timber burns up a startling 1,168 kilocalories per hour, and splitting logs with an ax takes care of another 444. By contrast, an hour spent watching television gets rid of 74 kilocalories; 281 kilocalories disappear while washing the floor, 405 skiing, 481 clearing snow, 510 playing soccer, 814 in a tough aerobic workout, and, finally, 1,213 in running fast—the only thing that beats lugging timber.

In other words: A Mars bar and a Coke are not going to be enough if you intend to spend the day working in the forest. What the woodcutter needs is *food*, not a snack, and the concept of the "lumberjack's breakfast" has real historical roots. Before the era of the chain saw, when giant trees were felled in deep snow, in the days of the horse, the ax, the two-handed saw, and the twelve-hour day, part of the worker's employment contract specified the amount of food he had a right to. Very often it was peas and pork in quantities that might make the modern-day nutritionist gape. A study carried out by a Swedish university of the food delivered to forestry workers in Norrland (the northern part of the country) revealed that each man consumed an astonishing ninety-three hundred kilocalories per day. The figure sounds impossibly high, given that the average office worker today uses only two thousand to three thousand kilocalories. The forestry workers were actually close to the limits of what people are capable of burning off, and not much below the highest figure registered: the twelve thousand kilocalories consumed by those on polar expeditions using skis and sleds.

Research shows that modern forestry workers using a chain saw burn off about six thousand kilocalories in the course of a workday. The daily requirement corresponds to that of soldiers engaged in military exercises; the dried rations issued to soldiers in the Norwegian army contain five thousand kilocalories. That is the sort of packed lunch you need in the forest, along with plenty of drinking water. Put simply, the right kind of food in the right quantities means greater safety.

As the mechanization of the forestry industry gathers pace, there is a vanishing number of foresters around who still possess adequate knowledge of the traditional techniques and skills developed to make life easier for foresters who did everything by hand. These old methods and practices, handed down through the generations and shaped by the hard necessities of the times, can still be a source of real inspiration. Before the advent of chiropractors, the work had to be performed intelligently and efficiently. One of the best methods developed in the undulating Norwegian forest terrain was the so-called *skrålihogst* (felling against the hill), in which two or three trees are felled, preferably still loosely attached to the stump, with their tops facing up the slope. Once trimmed, they can be used to roll other logs down, like a sort of primitive conveyor belt. Another very useful technique is the *fellebenk* (forester's workbench), in which a tree is felled in such a way that it remains attached to a stump thirty to thirty-five inches tall and is then trimmed. The next tree is felled so that it lands on top of the first one, at a comfortable and convenient height from the ground for the forester to work on it.

If you don't have lifting tongs, a good idea is to leave a branch stump about six to eight inches long that can be used as a handle.

Felling in boggy terrain or close to water is best done in winter when the ground or water is frozen. A good technique is to start work by a part of the river or lake that is frozen, and then work your way along the frozen bank, using the ice to move your logs along. Such skills and know-how have been kept alive by institutions like the Forestry Research Institute and the Norwegian Museum of Forestry, as well as by those who still use the old workhorse. The technologies of the past, including the trestle and the goat (a hinged sled for dragging timber), are as intelligent today as they ever were. The same is true of the wire pulley with block and tackle, and the only thing needed to make use of the most intelligent transportation means of all—floating the logs on water—is the simple long-handled cant hook.

The Cutting Season

The woodcutter who attunes himself to the ways of nature and the passing seasons will quickly find his reward. The annual growing cycle of trees means that the best time to fell them is in the winter or spring, well before the leaves have started to bud. The exact time depends on your whereabouts, and remember that the sap starts to rise some three to four weeks *before* the leaves appear. There is a minimum amount of moisture in the wood at this time, and it will dry a lot more quickly than it will otherwise. Ash can have a moisture content as low as 34 percent in the winter, while most other kinds of trees have in the region of 45–60 percent. The variation in moisture content over the year is not as great as is often claimed—usually it varies by about 10 percent, and in fact many trees have their lowest moisture content in late June.

What is most important about winter felling is that it gives a much longer drying season, and that fungus and mold can't establish themselves at temperatures below forty-one degrees Fahrenheit (five degrees Celsius). And in summer the sap in certain deciduous trees has a higher sugar and mineral content, exposing them to greater risk of fungal infection. A tradition in Norway is to fell the trees around Easter so that the timber can be dragged along the crusty surface of what remains of the winter's snow. In spring and summer there is more green to be dried out of the wood, and thus the trees weigh more and are more cumbersome to handle.

So late winter and early spring are the best times. But even if you're late getting started, or simply don't have the time until summer, don't despair. Even timber that is soaking wet in the early summer dries a lot more quickly than most people realize. It makes sense, however, to use a special technique called *bladtørk* or *syrefelling* (leaf-felling, or drying with the leaves on). It's common in Norway, though not widely used elsewhere. Here the trees are not limbed after felling; they are allowed, rather, to lie for some time with their branches, leaves, and tops intact. The leaves continue to grow and draw sustenance from the trunks, so moisture is transported out of the tree. (Oddly, the leaves will also sprout on deciduous trees that have been felled in the winter, before they have begun to bud.) When this method is used, the level of dampness falls quickly: The moisture content of a tree can fall from 50 percent to 35 percent during the first week. After a few more weeks, depending on the season and the size of the tree, the leaves will have withered. This does not mean that the wood is now ready for burning, as the moisture content will usually be still about 30 percent. The effect is most marked during the first few days, and the process stops completely once the leaves are dry and brown. At this point there is nothing more to be gained by leaving the trees lying, so it is vital to keep an eye on things and have the wood chopped, split, and stacked so that the drying can continue if the wood is to be ready for use the same winter. Evergreens will also continue to draw nutrition from the trunk if allowed to lie with their branches intact, but the process is so slow that there is little practical benefit to be had.

Another method of speeding up the drying is strip-barking, a process that involves stripping the bark in two or three places, which allows the logs to breathe and the moisture to escape. You need an ax, a bark spud, or a chain saw for this. Strip-barking and leaf-felling are techniques that work well together. For decades a requirement of Norwegian quality control was that logs intended for use as firewood be strip-barked in two to four strips, depending on the thickness of the tree, whereas deciduous trees leaf-felled before July 15 could be delivered unstripped if they had been lying with their leaves for a period of six weeks. Removing one or two strips of bark is particularly important when the wood is too thin to be split, otherwise it will take a long time to dry, because the exposed area at the ends of the logs is very small in comparison with the overall volume.

Another seasonal problem for the woodcutter in summer is the heat. The work is arduous, and when temperatures rise you will find yourself sweating profusely inside your heavy protective clothing. Besides the thirst and the headaches, the woodcutter will find himself tormented by the flies and midges that swarm from the branches when a tree has been felled. A consolation for these discomforts is that memories of such hard times make the heat from each log burning in the grate that little bit warmer.

Wood That Never Dries

So it is possible, but not ideal, to fell in the early summer. Felling any later in the year, however, may cause problems with the quality of the wood, even if you allow it two seasons in which to dry. Sometimes you find yourself with a load of firewood that just never seems to dry, no matter how long it is left. This is especially likely in the case of timber that has been felled in the late autumn and left lying in the forest for some time, or that has been stacked in an airless woodshed while still green. Even if drying conditions are improved later, the wood never seems to dry out properly. Why? Moisture is just moisture, isn't it?

Well, that's probably the question Norway's Forestry Research Institute asked itself back in 1964, at the outset of a two-year project on the drying of hardwood. Trees were cut down in each of the four seasons. Each batch was dried both on open ground and in thick forest, with the bark still on and with the bark stripped.

These trials, and subsequent research, showed that in deciduous trees for which the drying conditions are poor at the outset, the moisture content never falls to the level it does in timber that has been dried quickly. When the drying process does not begin and the moisture remains in the wood for a long time, the tree's natural degenerative processes commence. It doesn't necessarily become visibly fungal, but bacteria begin to establish themselves, and, in layman's terms, the enzymes and the slime form a film around the fibers. The transporting of moisture out of the tree is impeded, and even when conditions are improved, the drying never really gets going as it should. Even after a long period of drying the slime will remain in the wood and reduce its quality. The Sami people actually have their own word for poor-quality firewood like this: *tjásjsjallo,* literally, "full of water." It describes wood that has begun to rot and refuses to dry, even when stored under cover.

The negative consequences of such bacteria are twofold. First, the slime prevents the moisture from evaporating completely, so the moisture content

Don't let timber lie too long in the woods. A bad start to the drying process allows mold and dry rot to take hold and can reduce the quality of the wood despite good drying conditions later on. Pine and spruce can be left longer than deciduous trees, but remember that even these will fail to dry at all if they are left too long.

never falls to its optimal level. And second, the slime will continue to have a negative effect even in the stove, because it impedes contact with oxygen and leads to sluggish combustion. The rule is never to let timber lie wet in the forest with the bark still on it; get it back home and chop, split, and stack it. Birch and beech are particularly prone to fungal attack; spruce and pine, with their high resin content, are less so and can be left lying longer in the forest.

In Norway there are still those who observe a tradition of felling trees according to phases of the moon. Trees to be used as firewood should be felled when the moon is in its last quarter. This rule is noted in old documents from across Europe, and modern research by the Swiss scientist Ernst Zürcher confirms that trees do indeed have cycles of their own. The moisture content is the same throughout the month, but many tree species will shrink and swell slightly during a lunar cycle. However, contrary to traditional belief, the difference is most notable in the five to six days before and after the full moon. The wood is most dense four days after the full moon. However, the changes are so small that they have no practical effect on firewood—perhaps just a 4–5 percent difference in density, but the properties may affect critical use, like wood for musical instruments.

But even if the end result does not differ much, simply to observe this ancient tradition casts a uniquely esoteric light on firewood.

Which Trees Give the Best Wood?

Apart from thorough drying, the quality of a batch of firewood is measured by its heating value. This is essentially a figure indicating the density of the tree, measured in British thermal units per cord or in kilowatt-hours per kilogram. Hardwoods weigh more, and give more heat, than softwoods of the same volume. For example, an oak log will generate 60 percent more heat than a gray-alder log of the same size. But pound for pound, the alder and the oak produce the same amount of heat. Seen in this light, wood should ideally be sold by weight and moisture content, though in fact few countries do this. One problem that arises when wood is sold by weight is, of course, that the purchaser risks paying extra for the moisture in unseasoned wood. Recently the Norwegian Forest and Landscape Institute, the country's leading forestry management research center, began a project called From Cords to Kilowatts. The goal is the introduction of a practical standard for marking the heating value of wood; the plan at the moment is to label sacks of wood with the number of kilowatt-hours they contain. Once the yardstick of quality is the amount of potential heat the sack contains, the firewood market should open up for the sale of spruce and other lighter and less popular types of trees.

Tables recording the density of the different types of wood are plentiful (see page 187), but we should not be blinded by statistics. These are averages only, and as they are derived from trees growing throughout Norway, the local variations for each type of tree can be considerable. Generally speaking, we can say that conifers that have grown slowly in poor soil will be better and heavier than those that have shot up quickly in rich, damp, low-lying soil. The latter will be lighter and less compact, as can be seen from the wider spacing between the annual rings. The opposite is true for oak, hickory, ash, elm, and other so-called *ring-porous woods*, which will become harder and denser when growing fast. Forestry researchers in Norway have recorded large variations in density of conifers, particularly in spruce and pine. The average density is 840 pounds (380 kilograms) and 970 pounds (440 kilograms), respectively, per cubic meter. However, researchers have also come across examples of both that weigh between 660 pounds (300 kilograms) and 1,320 pounds (600 kilograms) per cubic meter—in other words, with the density typical for oak! Diffuse-porous trees like birch, alder, rowan, beech, aspen, and willow show less variation than conifers. Lowland birch grown in different parts of the country can vary by about 15 percent.

But the joys of burning wood do not reside in the decimal point or under the laboratory microscope. The hardest kinds of deciduous trees are traditionally the most sought after, but all types of trees can be used in the modern wood fire. Quick-growing, "bad" types of trees make perfect firewood for the chilly days of early and late winter, and burn well without building up a red-hot heat in the stove. The wood doesn't last long, but this can be remedied by burning it along with something harder, such as a thick log of beech, oak, or, best of all, elm. The log will continue to glow long after the lighter firewood has been consumed, and ensure that the fire doesn't go out. This is a trick much used by Norwegian farmers, who have to leave the house several times a day to work outdoors. As in life generally, people take what they can get, and many enthusiasts use a wide range of firewoods for the pleasure of the variation. They chop in a nearby wood and note how the different kinds of wood perform in the stove. They enjoy the snap and crackle of the conifers, savor the broad, flat flame of the aspen in an open fireplace, and in the evening will heat with the harder sorts of wood, because these last longer and leave a fine layer of embers ready for the next log.

The use of conifers in open fireplaces is not recommended because of the way sparks shoot out from the flames. The same goes if you are spending the night outdoors beside a fire. Especially toward nightfall, it is advisable to use wood from deciduous trees—in the cold your instinct is to creep ever closer to the flames, and the sparks will soon scorch holes in your sleeping bag and clothes. The great popularity of birch (in Norway more than twice as much birch is sold as spruce and pine) is actually a modern phenomenon. Records show that until the 1950s wood from coniferous trees sold in the same quantities as that from deciduous trees. During the Second World War it was spruce and pine that kept Norway warm, with consumption three times what it was for birch.

Kindling

To make sure the business of heating goes as smoothly as possible throughout the winter, it is best to start looking out for good kindling even during your first days in the forest. Hardwood is difficult to get going, and experienced woodsmen make sure to fell a certain amount of spruce and aspen as part of their harvest. The structure of growth in these makes it easy to chop them into fine sticks, and they make outstanding kindling.

Pinewood saturated with resin is also excellent for use as kindling and for flares, though it gives off a lot of soot. A good alternative is to start the fire with twigs from deciduous trees. In fact, birch twigs have a much higher heating value than wood from the trunk, and the same is true of birch bark. Both burn easily. To get the twigs manageable and compact, a lopper rig is practical. This consists of two rows of poles knocked into the ground, forming a column about six feet long and eight to twelve inches wide. Twigs and tops are forced down inside and cut with a chain saw. The bundles are removed, stored in heavy-duty sacks, and left out in the sun to dry. A structure such as this is also useful for cutting thin tree trunks and long branches. It can also be made of pallets and rough lumber.

Felling a four-hundred-year-old Douglas fir, 8 feet in diameter across the stump and 260 feet tall. Photo from Jonsered's 1985 catalog for the American market. The saw is a Jonsered 920 Super.

Its Own Weight in Kilowatts

Trees that are of modest size and easy to handle can provide a surprising amount of energy in return for the amount of work involved, and a few simple calculations will show just how efficient the forest actually is at harvesting sunlight. For years foresters have worked out the dimensions of trees from so-called volume tables, where the diameter of the tree 4 feet above the ground—or 4.5 feet, in the United States—is calculated in relation to its height. Depending on the type of tree, the factors involved vary somewhat, but if we use a standard calculation, a birch 50 feet high with a diameter of six inches at chest level gives a volume of 0.12 cubic meters, which is equivalent to 150 pounds (seventy kilograms) of normally dry wood. Burned in an oven that is 75 percent efficient, this tree will produce 225 kilowatt-hours. If the price of electricity is $0.13 per kilowatt-hour, then the tree is worth $30. At $0.21 per kilowatt-hour, its value rises to $48. A fully grown birch, one that is, for example, 80 feet high and ten inches in diameter, will yield half a cubic meter, 660 pounds (three hundred kilograms) of wood, and 960 kilowatt-hours, making its value $132 and $205, respectively, at the same electricity prices as above. Pine usually grows to a height of 100 feet, and large pines of a quality commonly reserved for use as timber in house building can easily cover two cubic meters each. Even at the lower heating value, a pine as large as this will generate 2,800 kilowatt-hours in the stove. Many Norwegian households that heat exclusively with firewood are able to get through a winter using the wood from seven or eight large pine trees.

If you were heretical enough to fell the world's largest (by volume) tree, the twenty-five-hundred-year-old General Sherman Tree, in Sequoia National Park in California, it would keep you in fuel for years as you sat afterward contemplating the enormity of your crime. The Sherman Tree is 275 feet tall, and its maximum diameter is 36.5 feet. The volume has been calculated at 1,487 cubic meters.

Many of the most common types of trees can live to be several centuries old, and one of the oldest known trees in the world grows in Sweden, on a mountain heath in Dalarna, near Trysil. It is a small spruce, 9,550 years old. More properly, this is the age of the roots, and the trunk itself is a mere 600 years old. Botanically speaking, however, it is one and the same tree that has attained such an enormous old age. Trees that have reached the greatest age in terms of one and the same trunk are, once again, the California sequoias. Several still-living examples are more than 3,000 years old.

Norway's tallest tree, measured by volume, is the Sitka spruce. The largest specimens in this country can grow to a height of 150 feet and have volumes in the region of twenty-three cubic meters, the equivalent of more than thirty normal adult pine trees.

Permanent Green Energy

An interesting thought experiment is to try to calculate the size of forest that would be required to meet the energy needs of a single household on a permanent basis. An exact figure isn't possible, of course—to quote the first sentence in the old *Norwegian Forestry Handbook*: "All measurements contain a margin of error."

In our case the variables are many. The amount of wood needed will depend on the size of the house, the sensitivity to cold of those living within, the quality of the insulation in the house, the type of stove used, and, not least, the variations in annual temperature. The volume obtainable from the forest will depend on the type of tree growing there, the quality of the site, and annual variations in rainfall and temperature.

Let us, then, in all due modesty, make the attempt. It is usual to calculate that an average detached house in Østlandet, in eastern Norway, a cold region of the country with below-zero-Celsius temperatures from November to March, will consume twenty-four thousand kilowatt-hours of energy annually. This includes the use of lighting, heating, hot water, and sundry electrical devices. Even households that make a real effort to heat with wood could not count on more than half the energy—twelve thousand kilowatt-hours—coming from wood. Anything above that would require a complete rebuilding of the house, or the use of wood-fired central-heating boilers to circulate hot water and preheat the hot-water tank.

A modern living-room stove with 75 percent efficiency will provide twelve thousand kilowatt-hours from about 6,600 pounds (three thousand kilograms) of dry wood. (Wood will of course never be so dry, but for the purposes of comparing energy, it is most convenient to reckon with zero percent moisture.)

The starting point for the calculation is simple: The wood has to grow by at least the same amount in one year as we consume in one year. The potential for growth is expressed in terms of the site-quality class of the wood, a well-documented quantity in many countries where site-quality class is used for tax-assessment purposes on agricultural land.

In Norway it is usual to reckon that an unmanaged birch wood of average site-quality class has a mean annual growth of about one metric ton per acre. In the more clement southerly regions the figure can rise to more than two metric tons per acre, as is the case in the Danish beech forests, for example, and higher still if the forest is intensively managed. The calculation shows that each household would require between 1.5 and 3 acres of forest to provide it with twelve thousand kilowatt-hours per year.

But as we shall see later, certain kinds of trees can have a mean annual growth of as much as 3,300 pounds (fifteen hundred kilograms) per acre, and in that case the required amount of "garden forest" would be just half an acre. It would, however, require close management and a lot of active fertilization.

From an environmental point of view we might squeeze the calculation at both ends by managing the wood efficiently, so that it produces more and absorbs a lot of carbon dioxide in the growing phase, and by using modern, energy-saving devices in the house, such as a boiler, or connecting the stove to the hot-water tank. Wood is the simplest form of bioenergy there is, and the view of many Norwegians is that the environment is better served by using firewood than by spending money on complex strategies aimed at reducing a trivial amount of electricity. Some people have pointed out that if we invested as much in the development of bioenergy as we do in efficient energy use, we would soon solve all of Norway's energy problems. In the larger perspective, there are good arguments for allowing a "crude" source of energy to take over all our heating requirements, reserving the precision energy of electricity for the running of the more advanced electrical equipment in the home.

Coppicing

An ancient way of stimulating the growth of trees, and one that enthusiasts in Norway have again started to show an interest in, is to exploit the ability of certain deciduous trees (including birch, ash, chestnut, hazel, oak, poplar, and willow) to regenerate from stumps—a process known as coppicing. What happens is that the young tree makes use of the root system of the previous generation and grows more rapidly than it would if it had to establish its own roots. Coppicing gives it a kind of straw that reaches directly down into the source of nutrition, and by the fifth year of its growth one of these tree-stump saplings can be twice the height of a tree grown from seed.

When this tree, in turn, is felled, the cycle begins anew, with the root increasing in strength each time. Coppiced birch stumps, known as "stools," can nourish new trees for up to two hundred years before rotting, and oak can be rejuvenated almost indefinitely—in Great Britain there are oak trees growing from root systems that are more than two thousand years old.

Coppicing was known from earliest times in Norway but never widely practiced. Between 1600 and 1850, oak forests in Norway were almost completely wiped out by the needs of shipbuilding, especially after 1632, by which time Christian IV, king of Denmark and Norway, had to look elsewhere, having used up all his native Danish oak as timber for his large new navy. Other types of deciduous trees had little commercial value, and there was nothing to be gained from investing a lot of hard labor in forest management. But in recent years coppicing has been practiced as part of an energy-crops scheme aimed at producing environmentally friendly fuel. The mean annual growth turns out to be between three and five times higher than it is in a normal deciduous forest. Trees that grow quickly during the first years have been exploited, especially willow and poplar. In Sweden, coppices have yielded between 5 and 6 metric tons of dry matter per acre annually, and birch 2.5 metric tons per acre. The trees are felled before they become "lazy," that is, before the slower rate of adult growth establishes itself.

Wood from trees pressured into growth this way is naturally less dense than that from trees grown at the normal rate; but as we have seen, pound for pound, different types of trees generate the same amount of heat. Energy-crop forests are intended primarily for large wood-chip boilers, but it is perfectly possible to turn an ordinary deciduous wood into a highly productive firewood plantation.

The first step is to clear the wood of windthrow and any unsuitable types of trees. Next a clearing saw is used to thin the wood until the distance between each tree is about five feet—two arm's lengths is usually reckoned to be adequate for birch and ash, though more slender types of trees can stand even closer together. The wood is then divided into sections—for example, as many sections as the age of the trees when they are due to be felled. For poplars intended for use as wood-chip fuel, this might be as little as five years, so that

would mean five sections. Birch can have a rotation period of fifteen to twenty years and more. But the division is essentially a theoretical exercise; the primary aim of a firewood plantation is to get the trees growing from the stools again rapidly.

This is done by clear-cutting an area each year, though not directly adjacent to an area cut the year before, since too much space would mean the trees would not have to stretch for light and consequently would not grow as tall as the trees in other areas. It is also important that the clear-cut area be sheltered from wind. The work should be done in the winter, at the very latest before the sap starts to rise, otherwise the stools will be exhausted. The optimal height for the stools varies from one type of tree to another, and it's best to consult a local expert before proceeding. Low stools about four inches off the ground, with a slightly angled cut, give the most vigorous shoots for birch. Ash stools should be high, and hazel very low.

The shoots will begin growing on the stools the following year, often in large sprays that need to be systematically thinned, and this becomes the annual pattern of the work. In time the wood will grow evenly and densely, with trees reaching the same height in each section. Real enthusiasts will also take a tip from the environmental gardener's handbook and fertilize with ash residue from burning because, apart from water and dry matter, wood ash contains all the elements that were present in the tree, including iron, potassium, phosphorus, calcium, and magnesium. Trees have no objection to a little extra nutrition. Ideally the ash should be spread on snow or mixed with water so that the rich supplement does not come as a shock to the system.

Trees can also take up heavy metals such as cadmium, lead, copper, and zinc from the earth. These will gather in the ash and make it unsuitable for use as a fertilizer, though it is a useful property in other ways: In Kågeröd, in Sweden, where the local government uses the town's effluent to fertilize energy-crop forests, the growth rate of the trees increased threefold, so the forest functions as both a sewage-purification plant and a source of energy. The ash is treated as hazardous waste. With the expenditure of so much effort, there can be no complaints about an inadequate exploitation of the life cycle. Ash from the trees cut down the previous spring either fertilizes the shoots that appear on the stumps the following year or cleans the forest bed.

Of course there is no need to be quite as ambitious as all this. A forest that is well looked after and productive has a dignity and a calm that is all its own. And a felling cycle of forty rather than fifteen years brings a sense of perspective

OPPOSITE Even an old birch stool can sustain two new young trees. Using the old tree's root system, they grow rapidly. Oak stools up to two thousand years old are still able to foster new trees.

to the passage of time that all the owners of forests learn to value. Things don't happen overnight in a woodlot, and you may well be thinning on top yourself and thinking about your retirement benefits before you begin to enjoy the fruits of your labors. Usually it is the next generation that reaps the real benefits. But felling trees in a well-managed wood is almost a ritual, and in terms of efficiency and the amount of wood produced, the returns will be a lot higher than in an unmanaged wild wood.

Some Different Kinds of Trees and Their Uses

This book deals with the kinds of trees most commonly found in Scandinavia, but all sorts of other types of trees found throughout the world make excellent firewood. In North America, in addition to the varieties already mentioned, hickory, locust, and ironwood are particularly prized. In most agricultural and settler cultures, hardwoods are usually regarded as "the best there is," because houses in the past were so poorly insulated and the stoves produced little heat.

But bear in mind that, pound for pound, all wood gives the same amount of heat, and it is by no means certain that chunky hickory logs are the ideal fuel for a modern and well-insulated home. Continual heating with wood can make a house too hot and the air supply to the stove has to be choked off. Better to have a mixed stack of hardwoods and softwoods that can be used according to the outside temperature. In periods of mild weather softer kinds of wood are perfect. Such fuel can be burned intensely without making the house too hot, so that there is very little pollution from the stove.

In the larger perspective it is important to fell the types of trees that grow quickly and create an overgrown landscape. Almost all trees can be used as firewood, and once dry, all of them will burn. In North America, however, there is a small group of trees that give off poisonous gases when heated. These include the poison sumac and the manchineel. They are comparatively rare, but if in doubt, consult a local expert. When you assemble a stock of wood, the individual tree itself is probably more important than the degree of hardness. For example, a tree with few branches and a single long trunk will be much easier to fell and split than a maple tree, most of whose volume is in the form of branches that are more or less crooked.

Beech

The beech is Denmark's national tree. It is also the most common tree and, as such, it is widely used as firewood. It has the highest heating value of all trees common in Scandinavia. It is comparatively rare in Norway, and many Norwegian woodburners might wish it would emigrate northward. Beech grows slowly, but can reach four hundred years of age. It can grow to a height of 130 feet and be as much as 5 feet in diameter. On account of its fine and even texture, beech has been used for centuries by furniture makers; it can be steamed and bent and turned on a lathe. Given beech's longevity, it seems odd that it has never given rise to the sort of myths and legends that often attach to other long-lived trees. It is curious, however, that in many Germanic languages and in Russian, the word for "beech tree" is the same as or related to the word for "book" or the word for "letter." A persuasive explanation is that the words can be traced to the days when people wrote on tablets, and beech, with its smooth texture, made an excellent surface on which to inscribe letters. The bark is dry and firm, almost like the skin of an elephant, and the wood is clean to work with. Its hardness means that a sharp saw is needed, but it is easy enough to split when it is fresh. Its average heating value is a whopping 3,032 kilowatt-hours per cubic meter.

Ash

In many cultures the ash is the tree of life. In Old Norse mythology, the branches of the immense ash Yggdrasil stretched around the world, and the first humans, Ask and Embla, were created from the ash and the elm, respectively. For a brief period in Norway in the 1970s a decoction made from ash became hugely popular. The return of an old folk belief in its curative properties saw it selling in huge quantities, even in drugstores, until the enthusiasm was quelled by the science of doctors and researchers, which revealed that the benefits of the wonder potion were somewhere between modest and none.

The wood is very tough and strong, and for centuries ash has been used to build the framework of carriages and wheelbarrows. Even today, the frame of the classic Morgan car, manufactured in the United Kingdom, is still made of ash. Ash has always been a popular firewood, not least because it contains less moisture than other kinds of trees. Newly felled ash has so little moisture that, even in winter, it will burn without seasoning, though not very well. The quality is noted in the old English saying "Seer or green, it is fit for a queen"; like all other wood, it is best, of course, when properly dried. It splits easily.

Ash dieback, a fungal disease, currently presents a serious threat to the ash forests in many European countries. The threat has wider repercussions, for ash allows the passage of a lot of light through its leaves, which in turn encourages a rich and varied flora around its base. If the ash dies the ecological damage will be felt down here as well.

Ash regenerates from the stool and it is therefore ideally suited for coppicing. The trunk grows fairly high before it bifurcates, making it easy to fell and split despite its naturally slightly twisting shape. Its density is excellent: 1,200 pounds (550 kilograms) per cubic meter.

Maple

The maple is a lively, handsome tree, especially in the autumn, when it drapes itself in a vivid orange. Unlike the leaves of most other deciduous trees, those of the maple produce a new coloring as winter approaches. The maple is the national tree of Canada, and its highly characteristic leaf features on the national flag. The wood is bright and glistening and is much used in carpentry and in the making of musical instruments, especially violins. The frame of a Steinway piano, which also contains six other kinds of wood, is made from the sugar maple, which is a good deal harder than ordinary maple. The sugar maple is also the source of the syrup familiar on the breakfast tables of North America as a topping for pancakes, being thicker than the sap from the ordinary maple.

The maple is quite robust and can survive salt and moderately polluted surroundings. It can grow to a height of almost one hundred feet. Its crown is tight, so that the floor beneath is almost permanently in shade. It branches low on the trunk, and trees that grow in the wild will have a large, round crown. Using it as firewood can be a time-consuming business, because much of the work involves cutting up the thick branches. There can be no complaint about its heating value: 2,820 kilowatt-hours per cubic meter.

Birch

Birch, queen of the Norwegian forests, enjoys a richly deserved reputation as firewood, to the extent that it overshadows other kinds of good firewood, and many Norwegians labor under the impression that it is the only option.

But its status as our national firewood is well founded: There is a lot of it (74 percent of Norwegian deciduous forest is birch), and it grows tall and straight. The exception is the mountain birch, which can be twisted, making it difficult to feed into smaller stoves. But birch growing in the bottoms of valleys and in lowland sites will, when grown in crowded conditions, have long sections of knotless trunk. The timber is quite hard (so hard, in fact, that the fuselage of the Mosquito fighter planes flown during the Second World War was made of birch plywood), and in its long, unblemished form is much in demand among furniture makers.

The so-called Royal Birch, in Molde, in western Norway, was long one of our most celebrated trees. A famous photograph of King Haakon VII and his son, Crown Prince Olav, was taken there in April 1940, soon after the

Chopping both heavy and lighter wood, and splitting it in varying thicknesses, will give you firewood for every occasion, from the slightly chilly to the freezing cold.

German occupation of Norway. It offered important proof that the king had neither abdicated nor been killed—it showed he had, rather, successfully fled the advancing enemy forces—and the tree became a central image in Nordahl Grieg's poem "The King." It was vandalized several times after the war and was finally brought low by the New Year's Eve hurricane of 1992. Ten years earlier, however, King Olav V had planted a second Royal Birch on the same site. This too was damaged, but a third tree, planted by King Harald V in 1992, has been left unmolested, though its crown broke during the spring storm of 2011.

Wandering back into the woods again, we see how much easier and cleaner birch is to manage than the spruce and the pine. Limbing is easy and there are no needles or sap to gum up gloves and equipment. The erect structure makes the logs easy to split and the woodpile builds quickly, white and handsome.

In the stove too, birch wood is an exemplary performer. Its heating value is high, and it doesn't spit glowing sparks out into the room. And as though all this weren't enough, the bark burns easily and is a great help during the kindling phase, and burned birch wood forms a dense bed of glowing embers. But birch has requirements of its own: It needs to dry quickly, and it degenerates rapidly if attacked by mold. Left to lie untreated on the ground, it soon rots.

Birch grows at its best until about fifty years of age, and rarely survives more than two hundred years. The downy birch can grow to a height of 65.5 feet, and the silver birch to 100 feet. The average density is 1,100 pounds (five hundred kilograms) of dry matter per cubic meter.

Spruce

Many woodsmen are inclined to turn up their noses at spruce because the heating energy tends to be low, and because it burns without forming a dense layer of hot ash. But it has a place in the woodburning economy. Recent figures show that the density of spruce can vary enormously, ranging from 660 pounds (three hundred kilograms) to 1,100 pounds (five hundred kilograms) per cubic meter. So although spruce can be less dense even than the gray alder, some specimens of Norwegian spruce can weigh as much as oak. Such old, slow-growing spruce is heavy and packed with energy.

Spruce burns readily and generates heat quickly, making it ideal for use in a cold house or cottage. No Norwegian stock of wood is complete without its supply of spruce for kindling. The kind of finely chopped wood used for baking in woodstoves and ovens is usually finely chopped spruce (or aspen), because it burns quickly and evenly. The temperature can be finely controlled by feeding the fire every three to five minutes. In former times spruce was often referred to as the "kitchen wood," whereas birch was the "living-room wood."

Spruce and other conifers have a tight internal structure, which causes them to spark and crackle as the pockets of resin explode, so the best use for them is in stoves or fireplaces with glass doors. A lot of people enjoy the sparkling and crackling for the theatrical glamour it brings to the fireplace, and in the past spruce was traditionally burned in the home on Christmas Eve.

Dealing with spruce out in the forest is not for the fainthearted, and a really bushy tree can have so many branches that the trunk is hardly visible. Pinewood itself is easy enough to split, but the tree often has large knots running up and down the length of the trunk, which makes it very difficult to split with an ax. A hydraulic splitter is one way to deal with this problem.

Due to the disproportionate demand for birch, firewood sellers often offer spruce at a price more reasonable than the difference in heating value would suggest. It means that spruce often gives better value for the money.

Spruce can grow to a height of 115 feet, and the Sitka spruce even higher than that. The highest conifer currently growing in Norway is 157.5 feet tall. The average density is 840 pounds (380 kilograms) per cubic meter, but, as we noted earlier, this can vary from 660 pounds (300 kilograms) to 1,100 (500 kilograms), and in exceptional cases be as high as 1,320 (600 kilograms).

Scotch Pine

This tree is the "big game" most frequently hunted in Norwegian forests. With an upright, rough trunk the pine is almost always managed as sawing timber. And fully grown pines are so huge and cumbersome to transport, cut, and split that the work demands a disproportionate amount of effort from the woodsman, not to mention the dangers involved in felling such huge trees. But the trunk is straight and easy to trim, and when properly dried, smaller pines make good firewood, with a good heating value. Green pine is almost impossible to burn. Even when added to an already blazing fire it is as resistant as asbestos, and its resistance endures. When building a fire in deep snow, the Sami of northern Sweden would place a layer of green pine logs at the bottom of the pit to prevent the fire from melting downward. Dry pine burns with a large flame and gives off a lot of light, and in the old days fires would be fed dry pine as a way of providing enough light to work by. Pine is a good way to get an open fire burning more brightly.

Pine trees with cuts or "wounds" in the bark will produce large quantities of resin to bathe the wound. This saturates the surrounding areas, which then become what is known as fatwood. Fatwood burns so well that it can be used in torches. The heating value of pinewood oil is twice that of pinewood itself, but in other trees the oil content is too small to have a noticeable effect on the total heating value. The density of normal pine is in the region of 970 pounds (440 kilograms) per cubic meter, though this can vary greatly, and slow-grown pine from poor soil can be very hard and dense. The chimney, stove, and pipe system should be swept regularly if you burn a lot of pine, because the oil content means that it leaves a lot of dry soot.

Oak

If you have access to an oak forest then you are very lucky indeed. In many cultures the oak forest has an almost mystical position, and oak is beyond question the tree that has played the most significant role in the development of Western civilization. One of the reasons for this is the characteristic way the oak tree forks as it grows. By splitting up a rough, crooked oak in the right way, carpenters were able to obtain a wide variety of different shapes and building materials of great strength. Much of the timber utilized in shipbuilding made use of the natural curves of the oak, and some of the surviving plans for the correct division of trees employed by the old shipbuilders even look like butchers' charts demonstrating the correct way to partition beef. One example: The area around the point at which the oak trunk divides in two provides an incredibly strong V shape, which was frequently used by shipbuilders to make ribs. One of the timber wonders of the world is the ceiling of Westminster Hall, in London, made six hundred years ago. Many of its arches follow the natural lines of the oaks from which they were cut.

Oak's most outstanding quality is its extreme hardness and toughness. This leaves it well equipped to tolerate the burdens imposed on it by the advances of civilization, whether in the form of warships, cathedrals, or gallows strong enough to bear the dangling weight of a condemned man. A study of old maps showing the distribution and spread of oak indicates that, at one time or another, large forests grew in the vicinity of almost all the great cities of Europe, North America, and Asia. Most Viking longships, including the celebrated Gokstad and Oseberg ships (now in the Viking Ship Museum, in Oslo), were built of oak.

Bearing all this in mind, it might seem almost blasphemy to assess the quality of oak as firewood. But here too it comes out very well. Its heating value is the highest of our most common varieties of trees, only slightly behind beech. It is surprisingly easy to split when unseasoned, and as it dries it gives off a scent reminiscent of honey. Sizable oak forests are found only in the far south of Norway, but the oak is spreading northward. It can reach two thousand years of age, and in Norway there are several examples that are more than a thousand years old. Some of the largest are almost ten feet in diameter. The average density is 1,200 pounds (550 kilograms) per cubic meter.

Aspen

Most of the world's matchsticks are made from aspen. It splits easily and burns with an even, controlled flame. Its heating value is not great, but the wood is easy to chop into thin sticks, making it ideal for kindling and for use in baking ovens. The seasoning process can sometimes be unpredictable, and even when it has been properly felled, split, and stacked, areas of moisture can persist inside some of the logs while others dry out completely. Researchers have not yet been able to provide a satisfactory explanation for the phenomenon.

A parasite commonly found on aspen is the fungus *Phellinus populicola*. The wartlike excretions are extremely hard and burn slowly. At one time it was common to use a few when making up the overnight fire, so that it would still be glowing in the morning.

Even in a gentle breeze the aspen rustles because its leaves are stiff and the stalks long and flat. This has given rise to legends all over the world, including the story that the aspen was condemned to shiver throughout eternity because it provided the timber for the cross on which Christ was crucified. Sadly for the story, the aspen wasn't native to the Holy Land at that time, nor is it mentioned in the Bible. Oak, by contrast, is mentioned a full twenty-six times.

In parts of Norway the aspen is a rare sight, but in the southern counties of Aust-Agder and Telemark it makes up a large part of the population of deciduous forest. The aspen settles readily on bare flats and grows very rapidly. It can easily reach a height of eighty to one hundred feet. Its average density is 880 pounds (four hundred kilograms) per cubic meter.

Other Kinds of Trees

Wood from the **elm** is strong and hard, at 1,190 pounds (540 kilograms) per cubic meter, but Dutch elm disease has made heavy inroads on the population. It is notoriously difficult to split with an ax, so a hydraulic splitter is much to be preferred. The stumps make excellent chopping blocks. **Willow** usually grows more or less alone so it is not much exploited as a regular source of firewood, but a log or two in the stack are a pleasant reminder of how useful this tree once was. The Latin word for "willow" is *salix*, and salicylic acid, the active ingredient in aspirin, was first extracted from willow bark. The supple wood can be used to make skis and barrel hoops, and no childhood is complete without its willow flute.

Rowan has always been highly prized in Norway because of the way the ash retains its heat for a long time in the stove. In other countries, however, there was a taboo against using rowan because it was valued as a protection against witches. Its usual density is 1,145 pounds (520 kilograms) per cubic meter, but it often grows as densely as beech. Rowan emits a strong odor as it dries. The **speckled alder** and the **black alder** are often overlooked because of their fairly low density—795 pounds (360 kilograms) and 970 pounds (440 kilograms), respectively, per cubic meter. But they burn well and provide a good return of heat per hectare. Under similar conditions, alder logs will dry better than other types of wood, all the way down to 8 percent moisture content, giving them in practice a higher heating value than other sorts of trees. The speckled alder in particular grows quickly and reaches its maximum height within twenty to twenty-five years. In former times the wood of the black alder was used for building dams and soling shoes. Alders love water and can often be found growing in dense thickets along riverbanks and in swampy terrain. In these settings it is best to wait until the ground is frozen before felling them. During drying the exposed ends of the alder take on a beautiful orange color that makes a feature of the whole stack.

OVERLEAF A Norwegian-made chain-saw bar from the early 1960s.

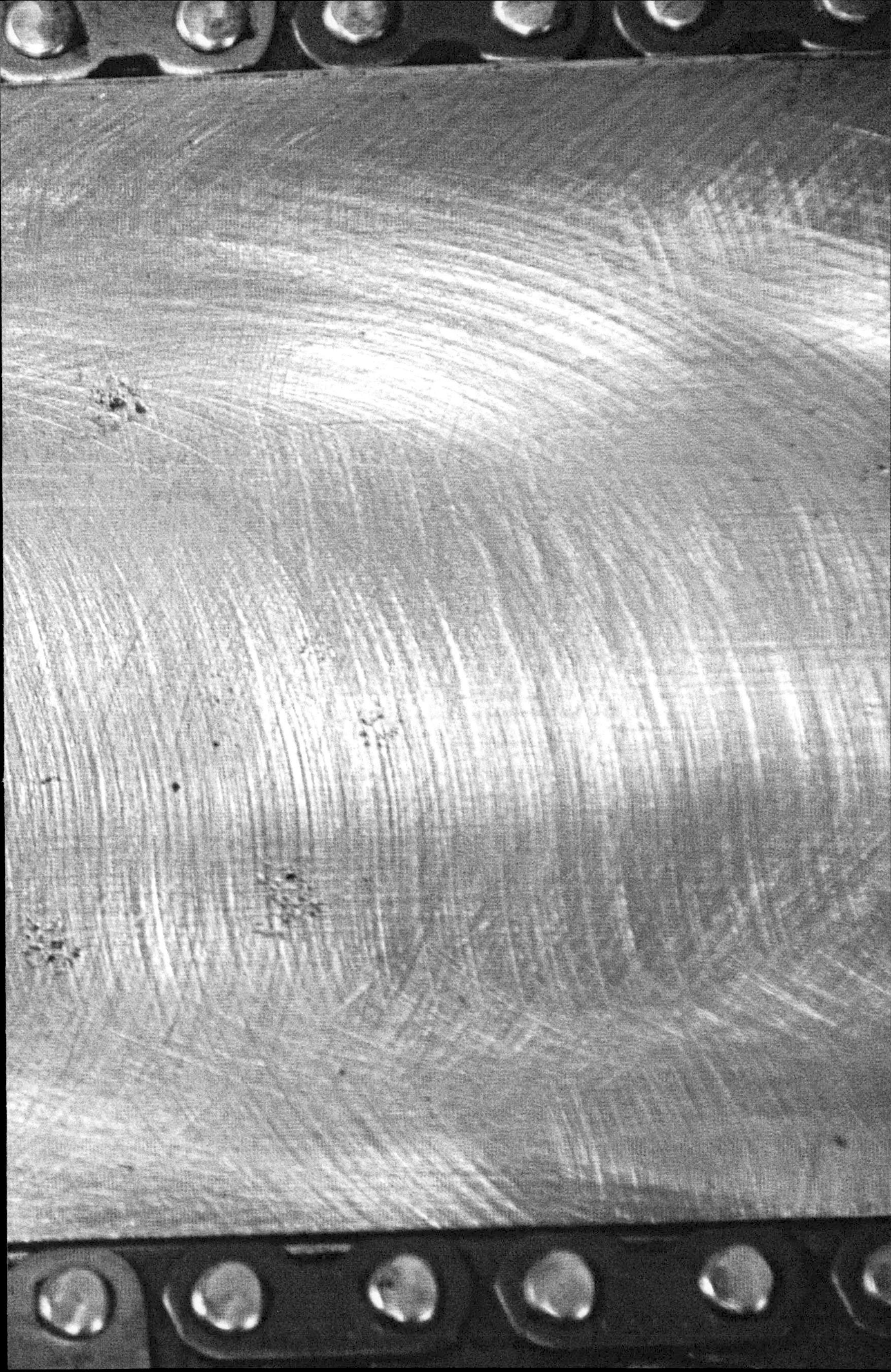

CHAPTER 3

THE TOOLS

CHAIN OIL

We've always used Jonsered on our farm.

–"Ola," on the forum of the Norwegian farming website gardsdrift.no

Hard work makes memories, memories that etch themselves into our tools. Every scratch on the saw, the ax, and the lifting tongs serves as a memento of days spent sweating in the forest. Many woodsmen develop a special relationship with their few and simple tools and the crusty patina of resin, time, and hard use that settles across them, for in this kind of work there is no room for inferior-quality equipment. Things must be able to survive a fall to the ground or being buried under the snow, and tolerate a coating of rust. These tools and the marks they bear from a lifetime of heavy duty ultimately become monuments to those who owned them.

At the heart of all tree felling is the chain saw. Few tools are capable of greater things per liter of fuel. Provided the chain is kept sharp, it will do whatever you ask of it. Come rain, snow, sleet, or hail, it will give you nine thousand revs, and nine thousand revs whether you hold it upward, downward, or sideways. Inside an hour it will transform an old pine tree into a batch of shaved and chopped logs, and with normal maintenance it will serve you year in and year out.

So this is not something to be bought at the garden center on a Saturday morning, with your children's fingers sticky from ice cream, your wife impatient to get on with the rest of her day, and the parking meter about to expire. Like a man's choice of hunting rifle, car, and sound system, the selection of a chain saw is something to linger over. There are catalogs to be studied, specifications to be compared, more catalogs to be plowed through in peace and quiet. Every decimal point relating to the horsepower and the vibration level of the handle must be minutely studied before the final choice is made. Only then has the groundwork been laid for the development of a genuine relationship with your chain saw. And this is not merely to make a fetish of the procedure. It really does make sense to become as familiar as possible with a tool this powerful and this efficient, because in clumsy and careless hands that same efficiency can prove fatal.

It's best to make your purchase on a weekday, when you can get time off work, the shop is quiet, and the salesman is in good shape. His expertise is every bit as crucial as any brand name, and small Norwegian towns tend to be divided into "Stihl towns," "Jonsered towns," or "Husqvarna towns," all according to the share of the market built up for these brands by the local dealer. For many years now these three have dominated the Norwegian chain-saw market. In former times many other good-quality makes were also available, but takeovers have shrunk the market. The German Dolmar chain saw, for example, is rarely encountered nowadays, even though it is still widely used by the Norwegian military.

So in the local cafés and bars the discussion is almost always about the relative merits of the German Stihl and the Swedish Jonsered or Husqvarna. The latter are available in a number of models that are more or less identical and differ only in choice of color and design. The attraction of the discussion is that it addresses itself to detail, and like other, closely related discussions (the merits of the Vauxhall versus those of the Ford, the Chevrolet contra the Dodge), it can go on indefinitely since the German and Swedish brands are, in fact, equally good.

So which saw to choose? The first rule is to find that friendly local dealer who can answer all your most pressing questions about the way the saw works and politely explain to you that, for your modest needs, in all probability the twenty-inch bar would be overkill. He will also sell you protective footwear, helmets, and any spare parts you may need, and he won't bat an eyelid when you ask him to demonstrate the right way to sharpen the chain.

Many manufacturers operate with three levels of quality: for the hobbyist (periodic use), the serious owner (regular use), and the professional (constant

PAGE 66 The bow saw is a time-honored tool, with a purity that really connects the user to the elements. The saw in this photo has a blade for dry wood—a special blade with asymmetric teeth is recommended for fresh wood, to keep it from dragging.

PREVIOUS LEFT A magnetic length gauge attached to the saw bar is a very handy tool. Small marks are made before the gauge is removed and equidistant logs are cut.

PREVIOUS RIGHT The first inches below the head of a splitting ax are prone to wear, and often the handle will break just at this point. This Gränsfors splitting ax has a protective metal collar, and a retrofit of a similar brace is recommended on other axes. Duct tape will also suffice.

use). Regardless of the choice you finally make, avoid at all costs buying a chain saw that is too large and heavy for your needs. Cutting wood with such a saw is akin to strapping on slalom skis for a cross-country ski trip. A good rule of thumb is that if the saw is never too small, then in actual fact it is probably too big.

Really good chain saws have a long life span, and secondhand saws can be bought at a reasonable price. First make sure that a skilled man with two-stroke oil under his fingernails has checked the chain brake, the sprocket, the clutch, and the other critical components. The beginner who goes straight from his office desk out into the woods with a borrowed and substandard machine that has never been properly looked after stands a good chance of adding to the accident and injury statistics.

The most common problem with old chain saws is that they are difficult to start, either cold or once they've reached working temperature. The problem often lies in the carburetor—the ignition system in saws produced after 1980 is almost always electronic and maintenance-free. Chain saws that are hard to start will often get a new lease on life once the carburetor has been given an overhaul. On most saws this is a simple device that will function perfectly once it has been cleaned, given a blow-through with compressed air (the air pump at a service station will do the job adequately), and had its gaskets and membranes replaced. The job can be done by the average mechanically minded man in the course of a single evening with good-enough light. Your dealer can supply you with an original repair kit, or one can be bought on eBay; the kit fits into a small envelope and won't cost you much more than a gallon of fuel.

It Must Be Sharp

Correct sharpening of the chain saw is more important than size. This is so crucial to good performance that Norwegian woodcutters use the term *skamfiling* (cowboy sharpened) to describe a badly treated chain. A sharp chain will produce large, square chips, and should be able to make its own way through without forcing the saw into the trunk. If roundish powder is what you're getting from your saw, then a sharpening is long overdue.

Each link on a chain saw has two teeth: a chisel and a raker. The chisel controls how far to let the raker go. A good routine is to sharpen the rakers every time you fill up with fuel and chain oil. You need a round file and a guide tool for this, and the aim should be to create the same angle on all the teeth. Sharpening should be done only on the forward motion, and it should be done the same number of times on each tooth—three will usually be enough. It is important that the chain be evenly sharpened, or it will snag and cause the saw to judder. Add a drop of red paint to the first tooth and you will know when you have completed the job. As the cutting teeth are gradually filed down, the chisels need to be filed down too, otherwise the chain won't bite into the tree no matter how sharp the cutting teeth are. The usual practice is to leave the chisels high when cutting harder types of wood or working in winter, and to file them down for softer types of wood, to speed up the cutting. A flat file and a separate guide tool are needed for the chisels. The guide tools sold by the various manufacturers can vary considerably in usefulness and it is a good idea to try another brand if sharpening proves difficult.

Equipment

Besides a good chain saw, a pair of chaps (protective trousers), a helmet, and boots are indispensable. **Chaps** have a lining that will unravel and jam the chain immediately if it penetrates the outer material. The **helmet** provides protection against flying chips and damage to your hearing. Models with a neck protector keep out irritating twigs and snow. The **boots** have metal toecaps to protect the toes. The **lifting tongs** are a superb application of the laws of physics. The tongs provide a firm grip at an angle to the log that greatly improves the working position as well as protecting the back. Any able-bodied person should be capable of using two of these at the same time, distributing the stress more evenly through the body as the logs are dragged forward. **Felling wedges** are very useful when chopping down larger trees—you wedge them into the cut so that the saw blade does not get stuck. But be warned, if the trees are big enough to require felling wedges, then this is probably no job for a chain-saw novice. **Combination cans** for chain saws have a small container for chain oil and a larger one for fuel. The fuel spout has an automatic shutoff mechanism and once you have tried one, you will not want to go back to fiddling about with a funnel and a jerrican. In the space between the two containers there is room to slip a **round file**, a **flat file**, and a **combination tool**. The combination tool has a wrench and a screwdriver for slackening and tightening the chain, and if necessary removing the spark plug. Some people like to lop off the twigs with a **forest ax**, but a large **billhook** is also very useful for limbing young deciduous trees. No matter how careful you are, sooner or later you are going to find yourself cutting into stone, and anyone who has seen the resultant shower of sparks will know that it is prudent to carry a **spare chain** in your knapsack. Forestry equipment is heavy and awkwardly shaped, and generally speaking, only a knapsack or a bag bought from an army surplus store will serve you for any length of time. Chain saws use lubricated fuel, but it is quite acceptable for even a macho man to use alkylate fuel. The exhaust fumes are marginally less irritating, and in fact most

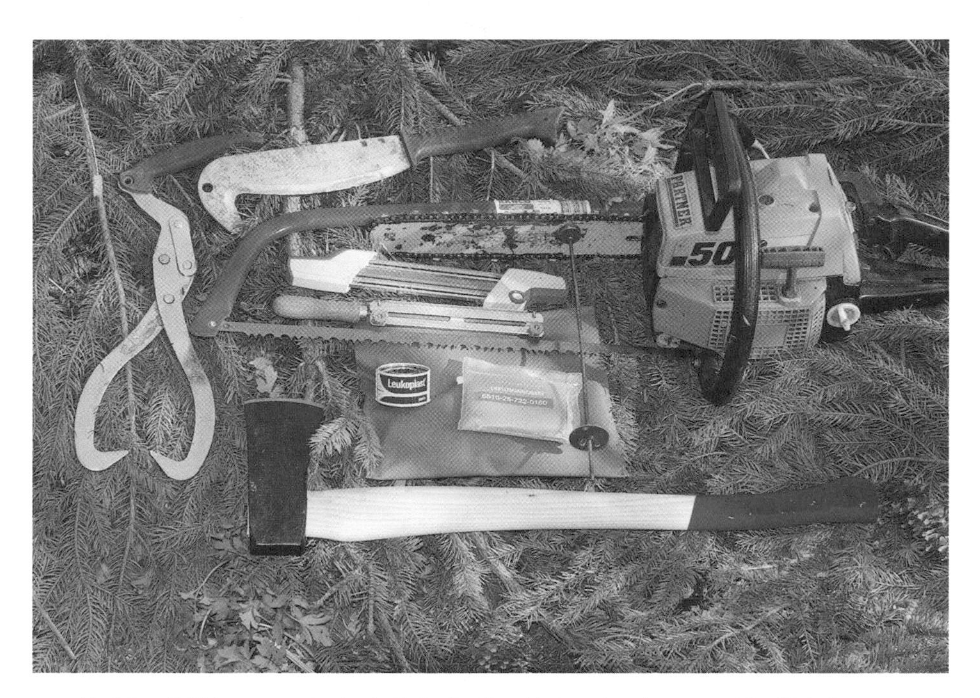

Useful stuff: lifting tongs, clearing knife for limbing thin branches, bow saw with a blade for green wood, files (a standard round file and a special type that cuts both the raker and the chisel at once), a length guide with a magnet attachment, a forest ax, and a well-provisioned first-aid kit, which should always include a roll of toilet paper.

saws will run slightly better on it than they do on two-stroke oil and ordinary fuel. A **first-aid kit** will contain whatever you need for large and small cuts, and it is a good idea to take along a **cell phone**. A **roll of toilet paper** will, as ever, often rescue the day.

The Most Common Types of Chain Saws

Stihl Made in Germany, this saw has a large fan base in Norway. The firm was started by the chain-saw pioneer Andreas Stihl (1896–1973), who took out a patent on the first functioning fuel-driven saw in 1929. In Germany it was known as the Stihlsche Baumfällmaschine Typ A. Stihl has stayed faithful to its characteristic white-and-orange casing, and good-quality German engineering means that older Stihl models are still frequently encountered in the forest.

Jonsered Jonsered saws stand for Swedish forestry culture. The Raket model appeared in 1954 and was, by contemporary standards, a very lightweight one-man chain saw. The company was named Jonsered after the town in Götaland

This Jonsered 590 produced in the 1980s has seen many good days of use.

where the saw was first manufactured; it was later bought by Electrolux, and now shares many of its parts and genes with a former competitor, Husqvarna. In Jonsered's marketing and choice of colors, it goes for a cool, youthful style. A line of cheaper models is made in Asia, with the remainder of the range being made in Sweden.

Husqvarna Husqvarna has been a familiar and much-loved brand for many years. Since 1959 its chain-saw casing has been in bright "Husqvarna orange," a color with a distinct tendency to fade over the years. In the mid-1960s the company launched the Husqvarna 65, one of the first saws that could be used both for limbing and for felling. Husqvarna is behind a large number of innovations in the field, and among the most recent is a chain brake that can be applied at both front and rear handles. Jonsered is the Volkswagen to Husqvarna's Audi, and the Husqvarna name is reserved for the very biggest saws. The largest has a 119cc motor.

Partner This company, originally AB Bergborrmaskiner, was making chain saws long before Jonsered and Husqvarna. It manufactured saws for professional use until the 1980s, when the name was bought up by Electrolux and thereafter used for a range of hobby models. Most models are now simple ones made in Asia and sold at hardware stores, but any woodcutter who turns up on-site with an old Partner saw is definitely a paid-up member of the club.

JoBu This Norwegian company was established in the dynamic years following the Second World War and was for a long time at the forefront of product development. A lot of people have a special relationship with the JoBu saw because the saws were developed and made in Norway, and because the quality was always first class. JoBu, too, was swallowed up by the Electrolux giant, and the factory in Drøbak closed down in 1983.

The Ax

We are indebted to our axes. Buildings, boats, furniture—all owe their existence to the genius of that remote ancestor of ours who first came up with the idea of attaching a sharp stone to a wooden handle and thereby greatly increased the power he could apply to cutting things, without any loss of precision. As weapon, tool, and status symbol, the ax has been one of our most important possessions for thousands of years. Today, more often than not, it lies forgotten in some corner of the garage, rusted and blunt, used only for chipping away ice or some other job for which it is pitifully overqualified.

A really good ax is worth etching your name on. Unlike a chain saw, an ax can never wear out; it can only be ruined. Naturally, you can buy a cheap ax made of scrap iron from some bargain-basement store, in the same way that you can, if you really want to, have cornflakes for dinner every day, or decide to stop changing the oil in your car. There's no need for Scandinavians to be bashful on this subject, because some of the world's finest axes are made up here—Hultafors, Fiskars, Øyo, Gränsfors, and Wetterlings axes, and the revolutionary new Vipukirves ax. There is a wide range of models available, and the selections complement each other. Categorical recommendations are difficult, and the ax one man swears by might make another man curse. Your own physique and the kind of wood you will be chopping should dictate the weight of the head and the length of the handle. Check in particular that the oval shape of the handle fits your grip comfortably.

Many people instinctively go for a crude, heavy model, but Newton's second law of motion applies in the world of woodcutting too, so that twice the speed quadruples the force of impact. A three-pound ax swung at forty-two feet per second has the same impact as a five-pound ax swung at thirty-two feet per second. Studies made by Hultafors show that three pounds is the most a person of average fitness can wield comfortably for any length of time, though of course the really heavy ax does have its place, along with extra muscle power, when wood proves particularly obstinate.

The variations in detail of an ax can cause surprisingly large differences in performance. A deep (long) axhead will give greater penetration into the wood, which is useful if there are a lot of small branches, but if you have the bad habit of twisting your wrist slightly as you strike, the point of impact will be increasingly crooked the deeper the axhead is. Long handles mean more speed, but it takes practice to ensure consistently good aim. A large end knob on the handle is very useful, preventing the ax from sliding out of your hand and giving the strike a greater impact. A curved handle gives improved ergonomics and acceleration as the upper hand slides down the handle, but twisting the wrist can compromise the accuracy of the strike. The old-fashioned felling axes had short heads and straight handles that gave equal balance whatever the angle of the strike. Most Scandinavian axes have fairly straight handles. The angle between handle and head has to be right for the intended purpose. Usually this means the head is set downward a few degrees relative to the longitudinal axis of the handle; otherwise it will not bite properly into the tree.

Old Steel

Now that the ax is no longer an essential tool of the trade for any group of workers, the quality of steel used in modern axes means that they are able to withstand a certain amount of ill-usage without too much damage to the blade. An ax from the 1970s or earlier may be superior in some ways to a modern one in that it will be easier to get a sharp edge on it, but it will be less resilient. Some of the old felling axes could be so fragile in the cold that woodcutters had to warm them up before use, and a distinction was usually made between summer and winter sharpening, with the winter sharpening leaving the edge with a slightly blunter angle. Nevertheless if you're lucky enough at a flea market to pick up an old axhead, minus shaft, of the quality of, for example, the legendary Norwegian Mustad no. 2 that was exported all over the world, attaching it to a new handle is an excellent way of giving old steel a new lease on life.

Never strike on the pole of the ax, and never use it as a hammer. The steel at the pole is not tempered, and the eye will lose its shape and the head will work loose. The best way to keep the blade in good condition is to use a flat

OPPOSITE An old method of splitting long logs was to chop them lying flat, hitting directly into the bark. A splitting ax with a narrow head is used, such as this one made by Wetterlings. If you don't have much space for splitting and drying, chopping the wood in long lengths this way is a practical alternative, even if the drying takes longer. More of the work can be done out in the forest, and dry cordwood is easy to transport.

The Finnish Vipukirves ax penetrates only about an inch into the wood. The rest of the energy in the blow turns the axhead sideways, breaking off the outer edge of the log.

file to rub away any irregularities, then sharpen it with a whetstone in water, or use a honing device. The honing devices sold by most manufacturers do a good job. Blacksmiths advise against high-speed bench grinders because within a few seconds the steel will get much too hot. At around four hundred degrees Fahrenheit (two hundred degrees Celsius), the tempering can start to break down and your ax will never be the same again. When the day's work is done, an ax with a wooden handle should be stored in a dry place at an even temperature. This will ensure that the wood does not shrink and start to come loose at the eye.

Different Kinds of Axes

The **forest ax** is used for limbing branches and cutting down small trees. It has a slim head and a curved blade that enables it to cut across the fibers of the wood. The head usually weighs about two pounds, and the handle is of medium length. The balance makes them ideally suited to chopping laterally. The cutting angle is sharp, often thirty degrees or less, and the edge has to be kept keen.

The **splitting ax** is intended not to cut, but to force a way down through the wood fibers, exerting a sideways pressure that causes the log to split. The head is wedge-shaped and feels slightly front heavy because it is balanced to cut directly downward. The head usually weighs between 3 pounds (1.3 kilograms) and 3.5 pounds (1.6 kilograms). For wood that is easy to chop, such as knot-free birch, a lightweight ax with a moderate wedge is perfect for the job. It does,

Seven good workmates from Scandinavian manufacturers: from left: Wetterlings Swedish forest ax, Vipukirves splitting ax, Øyo Maul, Gränsfors large splitting ax, Hultafors 3-pound (1.4-kilogram) splitting ax, Fiskars X17 splitting ax, and Øyo splitting ax.

however, have a tendency to get stuck, and wood that is tough, dry, or crooked is easier to split if the wedge is more pronounced. But in practice the shape of the head makes less difference than one might imagine when comparing splitting axes, because it takes more force to drive in a wide-angled wedge. One detail that is important is the angle of the final half inch of the edge. This is what makes the initial entry and starts to split the wood. Too narrow, and the ax may get stuck; too wide, and it could bounce back. Most manufacturers have settled on an angle of between thirty-two and thirty-five degrees, so it is important to keep it at that when sharpening the ax, though splitting axes do not need to be especially sharp. Gränsfors axes have a steel collar directly below the head that stops the handle from splintering, and other splitting axes can be protected by wrapping heavy-duty tape below the junction.

The **splitting maul**, or **sledge ax**, serves the same purpose as the splitting ax, but is better on very thick logs with a lot of knots, which need a really hard shock to split. The head can weigh between 4.5 pounds (2 kilograms) and 5.5 pounds (2.5 kilograms), so more power is needed to maintain your speed. This is a beast of a thing that needs a long, straight handle to get a good swing. The head is deep because it needs to penetrate deeper into tough wood before it will split. Although not easy to use, the splitting maul is a mighty tool to have at hand when chopping your way through the woodpile and encountering those obstinate logs that are crooked or knotty.

The **felling ax** is rarely seen today, but for centuries it has been the

forester's most important tool out in the woods. It is designed for cutting down trees—big ones too. Gränsfors and Wetterlings make two especially interesting types. Their full-blooded felling ax with a three-pound head is based on an old North American design. During the colonial era, the ax was the main tool used for all tree felling, and immigrants from all over the world brought their native smithing traditions to bear on the different types of trees they encountered in their new homeland. These two Swedish reproductions probably resemble most closely literary history's most celebrated ax—the one used by Thoreau at Walden.

With the passage of time the large felling ax has been rendered obsolete, but as a tool for chopping down trees it remains incomparable. It gives a real feeling of connection with forest life, and there is no more epic way of bringing down a much-loved or even much-hated tree (for example, a neighbor's overgrown maple) than with an American felling ax.

A Finnish Invention

You might suppose it would hardly be possible to change the way an ax does its job, but in 2005 a Finn named Heikki Kärnä was granted a patent for his **Vipukirves**, also known as the Leveraxe, which has found enthusiastic supporters the world over. This is a splitting ax, but instead of splitting the log by cutting it in half, it detaches its outer sections. The axhead looks like a large, curved chisel with a counterweight on the other side of the blade. The head is mounted sideways on the handle, making clever use of the laws of physics, for once the blade has penetrated the wood slightly, the axhead turns sideways and the counterweight acts like a crowbar. Unbelievers should do themselves a favor and try Kärnä's ax, because it works extremely well and, once the user is familiar with it, it turns out to be a very fast worker. It is at its best on large logs and will make short work of even very large logs, especially if they are straight-grained. Heavy knots are always a problem, but this ax will seldom get stuck. It does, however, require a technique all its own. Each blow has to land around the edges of the log, and your grip should slacken just before the blow lands, otherwise the important sideways movement will be impeded. The logs always fly to the left, and you can increase your speed by putting a car tire up on the block and making your way around it as you chop.

OPPOSITE Stig Erik Tangen, from Løten, with his open square woodpile, all of it split using an ax. This method of stacking is especially good for logs that are short or twisted. Unlike the closed square, this form is built using the same techniques as the round pile. In the background is a drying bin made with reinforcement mesh.

Ax Manufacturers

Øyo Brothers Founded in Geilo in 1882, this is the last remaining ax manufacturer in Norway. But the tradition goes back a long way. Archaeological finds show that iron was being extracted in this region two thousand years ago. Øyo axes are classics, and the owl symbol and the band of red on the handle are much beloved by Norwegians. All are hand-wrought, fairly light, and slender, and comfortable to work with over long periods of time. The splitting axes have a moderate wedge shape, and the special recess in the middle means that the ax is easy to extract from the wood if it becomes stuck. In 2012 the factory started production of a splitting maul having only the central part wedge shaped, giving it good penetration and powerful sideways pressure.

Hultafors The Swedish firm Hultafors manufactures a wide range of axes. Most are made at the old Hults factory, whose blacksmithing traditions date back to 1697. A common feature of their splitting axes is the slender, deep head, and they are among the few that can also be used for lengthwise splitting (see page 76). The splitting maul has a head weighing five pounds (2.5 kilograms) that is slender and deep, giving particularly efficient penetration on pine and spruce. The pole of the splitting maul is tempered and can be used on wedges.

Gränsfors This firm has done much to keep alive the tradition of the older types of axes. They also make throwing axes and battle-axes. Their handmade axes encourage a pride in ownership and convey a strong sense of individuality. Where even surfaces are not necessary, the finish is deliberately left rough. No two axes from this modest Swedish producer are alike. Each smith adds his initials to the axes he makes, giving them a personal touch. The attention to detail is exemplary, from the smell of linseed oil on the hickory handle to the string used to fasten the company's handsome *Ax Booklet* to the handle. The booklet includes a photograph of the smith who made your ax. The heads of both chopping and splitting maul are wedge shaped.

Wetterlings This company has been in operation since 1882. Their ten blacksmiths forge axes with the same loving care and attention to detail as their counterparts at Gränsfors. The splitting ax has a slender, deep head weighing three pounds (1.5 kilograms). The splitting maul, with its five-pound (2.5-kilogram) head, was made using the same slender profile until 2012, when the design was changed to a more pronounced wedge shape. Both have straight handles with a large end knob. Wetterlings makes outstanding forest axes. The smallest of these, known as the "Swedish forest ax," has a head weighing just two pounds (1 kilogram) and is used for limbing branches. The heavier, three-pound (1.5-kilogram) model is an out-and-out felling ax based on older American designs.

Fiskars Fiskars axes have a synthetic handle molded to the axhead, which is coated with Teflon. Functionality is the priority here, and even those traditionalists with a fondness for hickory handles and worn blade protectors of wizened leather will usually concede that these "high-tech" Fiskars axes are annoyingly good. The heads of the X25 and X27 splitting mauls weigh four pounds (1.8 kilograms), and the handles are long—28.5 inches and 36 inches, respectively. The pronounced end knob gives the ax a good swing.

The Miracle of Hydraulics

Even the most die-hard supporter of the traditional ax has to concede that at least one modern invention in the field is nothing short of a modern-day miracle: the hydraulic log splitter. These have been sold in vast numbers in recent years, for the simple fact that at the touch of a button they give you at least four tons of controlled pushing power. The even, powerful pressure of the blade is very effective on the kind of tough, stubborn wood that resists the shock of a conventional ax blow, and the splitters make short work of even the most twisted and knotty logs. The maximum log lengths that the most common models of splitter can handle are 14.5 inches and 20.5 inches; the real difference in quality between them lies in the speed, especially of the return motion. Inexpensive splitters can be sluggish performers, whereas the more expensive models are faster and often have an adjustable working length.

Other Tools

An **electric chain saw** is suitable for use when cutting logs in a built-up area. It is not as powerful as a fuel-driven saw, but makes considerably less noise, a sort of muted buzzing. The noise from a fuel-driven saw can be both irritating and disturbing for the neighbors. Many people have strong opinions on the subject of tree felling because it alters the appearance of a neighborhood, and the noise of someone working with a chain saw can quickly awaken fears that some such thing is going on. In certain parts of Norway, the law stipulates that fuel-driven chain saws may not be used between Saturday afternoon and Monday morning!

OVERLEAF Ole Kristian Kjelling and his wife, Zofia, with the Rossini sculpture stack.

10
ROSSiNi

Another handy tool is the **Smart-Splitter** made by the Swedish firm Agma. This is a simple slide hammer attached to the chopping block. The axhead is placed against the log and you raise the hammer and throw it down against the log. The device is easy on the back, and makes it possible to split even quite lengthy logs with a minimum of effort, provided the wood isn't too knotty. The same manufacturer also sells the **Smart-Holder**, a fine sawhorse that gives a good working position when cutting with a chain saw. Its pendulum lock has an adjustable jaw that can hold logs up to ten inches thick firmly in position. The price of **log saws** has gone down in recent years. These are efficient, but care is necessary when using them. Although modern designs with the blade enclosed have improved safety, any Norwegian country doctor will confirm that the log saw probably holds the record for the number of digits lost. The **splitting wedge** is a joker you can play for those long or extra-thick logs, or if an extra display of manhood is called for when wrestling with the most recalcitrant logs. The wedges come in two styles, helicoid and straight. The helicoid wedges are best for firewood because they screw their way in and exert a powerful sideways force without falling out, whereas the straight wedge is designed for timber to be used in construction. Fiskars makes a version with a synthetic head that prevents splintering and gives a gentler recoil.

A **length guide** is essential for the pedantically inclined. A good tool can be made by attaching a thread bar to a strong magnet. By fixing a large washer between two nuts on the bar, the guide can be made adjustable to any length. The magnet is then attached to the saw bar, so that evenly spaced saw marks can be cut on the log. The guide is removed before cutting. It is also possible to attach a similar rod behind the saw-bar mounting nut.

A good approximation of length can also be had by simply placing the saw bar against the log and measuring with your eye.

The **bow saw** is probably not the ideal tool to provide wood for a whole winter's burning, but the pleasures of using it include fine, quiet days in the forest with a minimum of cumbersome safety equipment to bother about. Many enjoy the workout this kind of "analogue" cutting provides (mainly, we might as well admit, ex-military types and gym teachers), and the bow saw is excellent physical recreation for those slack "office muscles" in the back and shoulders. This is a good standby too if the chain saw gets stuck. Bahco and G-Man (the G comes from Grorud, in Oslo, where there was a factory that at one time was a major producer of saw blades; the shortened form of the name is for the international market) make some of the best bow saws around. The most practical sizes are between twenty-four and thirty-six inches long. The blade itself is all-important, and it is worth noting that there are different models for use on unseasoned and on dry wood. On the former, every fourth

Arnold Flatebø, from Lindesnes, in southern Norway, works on his ring stacks. Despite the unique advantages of this form of stack, it is rare. The framing consists of two semicircles of channel steel bolted together top and bottom to form a retaining hoop five feet across. The hoop is raised on logs to keep it off the ground. Once the structure is full, a wire is used to bind the wood and the hoop is removed. The stacks can be lifted by crane.

or fifth tooth is extra broad, to rid the cut of the damp sawdust that would otherwise clog it up and cause the saw to drag. The blade for unseasoned wood is also recommended for use on frozen fresh wood.

No matter how pleasurable it may be, physical labor has its limitations, and as tools of the trade, most modern farmers and wood merchants choose machines more powerful than the human body. In many cases this will mean a **hydraulic splitter** connected to a tractor, or a semiautomatic wood processor. **Wood processors** cut and split at a remarkable pace, and with a small conveyor belt and hydraulic feeder the user is spared a great deal of heavy work. A lot of people dry the wood in large sacks on pallets. Once the wood is dry it is a simple matter to transport it from place to place using a tractor. The whole procedure is a real blessing for farmers with a number of large buildings to heat. Most commercial firewood in Norway has been through the wood processor, and this machine has made firewood production a growth industry in Norway for the past twenty years.

PIONEERS OF THE ELECTRIC SAW

One name is writ large in the history of the chain saw—the Norwegian firm JoBu, a telescoping of the names of its two founders, Trygve Johnsen and Gunnar Busk, who met in the Resistance during the Second World War. Johnsen owned a small sawmill and his friend Busk was a gunsmith (as it happens he was also the man behind the legendary Busk peep sight, which rapidly became popular with Norwegian shooting enthusiasts because adjustment changed the point of impact by precisely one centimeter for every one hundred meters from the target). After the war, in the summer of 1946, the Holmenkollen ski jump was being repaired and chain saws were used to clear the landing slope. Johnsen and Busk wandered along to have a look and saw a British chain saw being used. It was typical of its time, a monster of a thing weighing more than sixty-six pounds (thirty kilograms) that took two men to operate. The saw was unreliable, impractical, and unpopular with woodsmen. The two decided on their way back into Oslo that they would try to design a chain saw of their own and in due course bought a 98cc auxiliary motor for a pedal bike from a scrap merchant. Their fuel tank was made out of a Primus stove, and the frame for the prototype from a radiator pipe. The gunsmith had an agile and innovative mind, and in time many of his ideas gave real impetus to the development of chain-saw technology. Among other things, Busk introduced the direct-drive chain (one that has no link between the crankshaft and the chain) and the centrifugal clutch, which meant that the chain did not move when the motor was idling.

After a couple of years spent in product development, including experiments with the leftover engines of collapsible military motorcycles made specially to be dropped by parachute, the two men were ready to go into production with their first chain saw: the JoBu Senior. It weighed 37.5 pounds (seventeen kilograms) and had a 125cc motor that gave 4 horsepower per 4,000 revolutions. (A typical chain saw today weighs less than eleven pounds, or five kilograms, and has a 50cc engine that produces 3.5 horsepower per 9,000–11,000 revolutions.) For its time it was one of the most practical chain saws in the world, but in those days, right after the war, it was hard to get hold of engines and parts, and people were wary and reluctant to get involved. The two men mortgaged everything they owned, and before the first JoBu saw was in the shops they had debts of more than $150,000—this at a time when it cost a mere $0.04 to send a standard letter. But in time the faith of our pioneers was rewarded and in due course seven thousand of their Senior chain saws were produced.

Jobu Skogsredskaper A/S

DOKKA — TLF. 35

In the days immediately following the Second World War, when the telephone was tested in Norway, the more important owners were allotted low numbers. In the tiny settlement of Dokka, in southern Norway, the JoBu chain-saw company had telephone number 35 on the local exchange, eloquent testimony to its importance in the community. This is an advertisement in a local newspaper in 1976.

Although it may seem strange today, chain saws were regarded with suspicion at that time and there was much resistance to their use. Manufacturers had to work hard to persuade skeptical lumberjacks and public bodies of the importance of this new technology for the rationalization and modernization of forestry in Norway, and to achieve a turnover that would enable them to continue with product development.

There were quite a few colorful players in the early days of the chain-saw industry in the 1950s. The competition was hard and the business attracted people with a fiery temperament. One legendary character was John Svensson (alias Chain Saw Svensson), who imported saws made by the Canadian firm Beaver. He had been arrested and tortured during the war and for the rest of his life suffered pains in his arms and joints; when demonstrating the Beaver saws he always made a point of stressing how the vibrations that passed up through the handle brought a welcome relief to his aching joints.

Svensson was not a man to take professional disappointments lying down. On one occasion he was so annoyed when a visiting government delegation refused to let him demonstrate his chain saw to them that he felled five trees across the road to stop them from leaving. Criticized for his actions in a local newspaper, he turned up unannounced at the newspaper's offices, started up his Beaver chain saw, and cut the editor's desk in half. He then got into his car and set off for Oslo. After about an hour he had second thoughts and decided to go back. His return caused panic in the offices, but this time he had come only to apologize and offer to pay for a new desk.

In the early days the only marketing ploy that really worked was to travel around giving practical demonstrations of the efficiency of the saws.

Norsk Skogbruksmuseum

Elverum

LA DET NYE ÅR FÅ EN
FRISK START!

En frisk og rask start er også en av JO-BU TIGERs fordeler — blant mange andre. Snakk med noen av de tusener som allerede har denne sagen! De kan fortelle hvor fantastisk rask og kraftig den er, og hvilke store fordeler man har av Servoclutchen når sagen er under tung belastning. Spesialmotor, spesialkjeder, automatisk kjedesmøring og lav vekt — bare 10,5 kg — er andre fordeler JO-BU TIGER byr på. De vil fort merke hvordan denne direktedriftssagen gjør arbeidet lettere for Dem!
Og husk: **Ingen service er som JO-BU service.**

Veil. pris **kr. 1.590,—**

JO-BU SALGSKONTOR a/s

HOVEDKONTOR: HOLTEGT. 28, OSLO NV. - SENTRALB. 60 26 90 - BUTIKK OG UTSTILLING: KEYSERS GT. 1 - OSLO - TLF. 33 71 12

But there was a Wild West atmosphere about the business, and sabotage was not unknown. During one demonstration in Finland the delegation from JoBu found it advisable to sleep with knives on their bedside tables and the saws hidden beneath the beds.

The highly competitive atmosphere led to rapid advances in the technology. Busk and Johnsen maintained their pace as front-runners, and their legendary JoBu Junior model is now acknowledged as the first genuinely practical one-man saw. It weighed more than twenty-two pounds (ten kilograms) and was reckoned the best in the world in its weight class. The factory turned out more than forty thousand of these Junior saws.

On the early chain saws the carburetor had to be rotated manually when the saw was turned sideways for felling, until the appearance of the JoBu Tiger in 1960. For the first time, here was a saw that worked in whatever position it was held in. In keeping with the utilitarian spirit of the first postwar decade, the JoBu saw could be put to other uses, and among the extras available were an earth drill and a propeller mounted on a stem that meant the saws could double as outboard motors.

The success of JoBu was a Norwegian business fairy tale, and for a while the company was the world's largest producer of chain saws. It remained the market leader for several decades in Norway, and as recently as 1977 still had 250 sales outlets across the country. The first factory was in Oslo, before the company moved to Drøbak. Like Jonsered and Husqvarna, JoBu was soon bought up by Electrolux. In total there were thirty wholly Norwegian-made JoBu models. Some of the saws produced in the 1980s were indistinguishable from Jonsered and Husqvarna machines. The last genuinely Norwegian model appeared in 1980, and the factory in Drøbak finally closed down in 1983. Original JoBu saws are now collectors' items.

OPPOSITE JoBu treated itself to a full-color advertisement for the new Tiger model in the January 1962 issue of *Skogeieren*.

CHAPTER 4

THE CHOPPING BLOCK

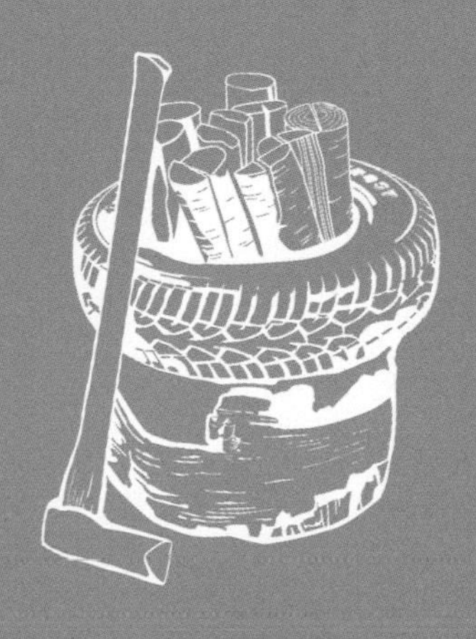

People love chopping wood. In this activity one immediately sees results.

—attributed to Albert Einstein

The chain saw has fallen silent and now your back is slowly recovering from the strain of dragging the heavy logs across to the trailer. So far, the work has largely been about trees, big and small; about two-stroke engines, limbing branches, loading and unloading, load straps and heavy tools. But now another interesting phase begins: the chopping and splitting that turns logs into firewood. Splitting and drying wood is a whole little science unto itself—and come winter it will mean the difference between good wood and bad wood.

Many people are at their most contemplative when addressing a chopping block. The work is a fine mixture of repetition and variation, and is often the first proper job undertaken outdoors after a long winter. Out come the axes and the log splitters and soon the whole countryside is abuzz with the sounds of the saws of retirees who feel like useful citizens again. There is a smell of fresh resin and sap in the air, and here is probably the right time to quote Hans Børli, from his book *With Axe and Lyre*, on the smell of the woodpile: "It is as though life itself passes by, barefoot, with dew in its hair. . . . When the veil finally starts to fall the scent of fresh wood is among the things that will linger longest in the memory."

Chopping firewood with an ax is one of the most *primitive* jobs left for a modern man to do—primitive in the sense that we do the same job in exactly the same way as our remote ancestors did. This is a chance to wield a heavy, handheld tool with all the strength that you possess. For a few blessed hours the simple but concentrated business of striking steady, rhythmical blows with a lump of forged steel banishes all the burdens of modern life. For the woodchopper cannot afford for a single moment to let his thoughts stray. The work requires your full and complete attention, and if it doesn't get it you might find the ax sticking out of your leg.

Chopping wood is an opportunity to use power—brute power, if you will—to triumph over something. You say you'll be damned if that stubborn, knotty pine log is going to get the better of you. Months later, with a nod mingling self-satisfaction and admiration for the log's resistance, you savor with particular pleasure the warmth that very same log dispenses as it slowly turns to ash in your fire.

Yes, it is an effort—but wasn't it always thus, that sweetness comes after difficulties? Is it not perhaps especially refreshing, in this modern day and age, to be able to engage in an ascetic activity that has been practiced in the same way since the dawn of time? Demanding manual labor such as this brings a kind of peace that is rarely found in other modern occupations. In life, whether it be at work or at home, we are always able to manage just that little bit more. Perhaps things will turn out even better if you continue working until eight that day, and check your emails after the kids have gone to bed. Almost everything will look better if you manage a bit of work on the weekend too. In our private lives we might often want to be more considerate, more active with the kids, more open about the things we normally feel uneasy talking about.

Physical work creates a kind of spiritual peace. Once a log is split it stays split. You can't change the split, or improve it. The frustrations of the day disappear into the wood, and from there into the stove. One of firewood's most attractive qualities is that it burns up and disappears. No committee will ever study it, nor will it be compared with another, competing log. Sooner or later, in the course of the winter, all the logs that have been badly chopped and clumsily split will end up in the flames too, and the heat they provide will be indistinguishable from the heat provided by the perfect logs—and isn't there an added spice to the pleasure of burning that particularly obstinate pine root?

Anne-Berit Tuft, a union leader from Oslo, recalls a time when her firm was involved in negotiations aimed at cutting the workforce. In the middle

PAGE 92 A low chopping block will give high speed and good impact to the blow. Here, a chopping block just 30 centimeters (12 inches) high is used with an Øyo ax.

PREVIOUS LEFT Pine stacked in the round by Ruben Knutsen, from Hamar.

PREVIOUS RIGHT Most wood species are easiest to split when fresh—especially important with large logs. Here, an oak grown near an old World War I battlefield in Somme, France, is cut for firewood.

of the negotiations she took a break and traveled out to her country cottage. Frustrated and depressed, she had just one thought in her mind: to chop wood. The most stubborn and recalcitrant logs in the pile had no chance against all that pent-up energy, and within a few days they were all split and stacked, and Tuft was back in the capital, brimming with renewed energy and raring to go.

The "Wood Age"

Elderly Scandinavian men with a passion for firewood are often told that they have entered something called the "wood age," or that they have been bitten by something called the "wood bug." The anthropology surrounding a passionate concern for firewood has not been the subject of much study in Norway, but research carried out by the Swedish University of Agricultural Sciences in 2007 appeared to confirm that a "wood age" does indeed exist as a distinct and measurable state. Nine hundred families living in Sweden were studied—the criterion was that all used woodburning stoves—and the results were unequivocal: It is men more than sixty years of age who spend the most time dealing with wood. Only 29 percent of the women in the study took any interest in firewood. Their average age was slightly lower than that of the men, and in the main their involvement was confined to stacking wood.

Working methods proved to be similar regardless of age. The average woodsman felled with his chain saw and split with his hydraulic splitter, though about 10 percent still used a bow saw and 21 percent split the logs with an ax. People tended to hang on to their tools and those taking part in the survey had owned their tools for an average of thirteen years. In the case of axes the average was an amazing fifteen years. More than a third of the nine hundred involved in the study astounded the researchers by saying that they would not part with their tools, not even if offered new and better tools free of charge.

One of the major findings of the study was that an interest in firewood can be related to a man's view of himself as a provider. Young men who had not yet started a family had almost no interest in wood whatsoever; by the time they reached their thirties and forties they were spending more and more time chopping wood; and once they had started they did not stop until they were past seventy. The interest climaxed once they reached retirement age, with the men spending an average of ninety-eight working hours per year on firewood-related activity. The finding seems natural: Retirees have more time on their hands, and they need an arena in which to continue doing useful work.

Working with wood can be a godsend when the ability to do more complex work begins to decline. "Wood saved my father when he began to suffer from dementia," is how this story from the Kongsvinger area, in Hedmark, begins.

The old man's grasp of the world around him was failing rapidly, but he was determined to continue preparing wood for the fire. The family was afraid he might harm himself with the chain saw, and deposited it for safekeeping on a neighboring farm. Each morning after breakfast the old man would come into the kitchen and complain that he couldn't find his chain saw. The family told him it had been sent away for servicing and would be back soon. "Well, then," he'd say, "I suppose it'll just have to be the bow saw again today. Never mind, that'll do the job." And off he went to start on a long—but safe—working day. Let us hope that in the midst of his isolated world he knew that "wordless joy" so well conjured by Hans Børli in *With Axe and Lyre*:

> If you are master of the art of setting your gear up properly then there is a special joy in seeing the way that sharp steel eats its way through the logs. The smell of resin and fresh wood, the sight of the smooth, clean cut made by a newly sharpened axe blade—such things can fill a man with a wordless joy and put him in touch with the essential joy of all physical labor: In his own hands he weighs the feel of life itself.

The Chopping Block

The chopping block is the woodcutter's own personal monument. The more slashes, cuts, and chips, the more proudly does it stand in the yard. A simple and straightforward item it may be, but great care should be taken in ensuring it is the right size. It is the ax's partner, and no ax can give its best if the block is unsteady or not really suited to the purpose.

The most important thing is that it be broad, so that it stands firmly, and that the ground beneath does not yield to the working surface and leak energy from the ax blow. Ideally it should be set on a rock base—or at least somewhere where the ground is hard.

The height is surprisingly significant. Most people look to protect their backs by going for a high block, but it is a contentious issue and any attempt to open a discussion on the subject quickly shows that the proper dimensions of a chopping block are a highly subjective matter. The tradition in Norway is to use a tall chopping block, but there are undeniable flaws in this practice. Too high, and the ax will fail to reach full speed by the time the arc of the swing meets the log. A blow that hits high at two o'clock in the swing will strike with less force than one that hits at four o'clock—as we noted earlier (on page 75), doubling the speed of the swing gives a fourfold increase on impact. The ax works best when the edge strikes the wood at an angle of almost ninety degrees, so the height of the block has to be appropriate for the height of the wood being

Fastening a car tire to your chopping block can save you a lot of work and a strained back. You don't need to be constantly bending down to pick up logs that fall from the block. It also makes it easier to split small or long logs. Chains and straps are an alternative when dealing with bigger logs.

chopped, the length of the ax handle, and the woodcutter's own height. It is tempting to propose some kind of formula here, but instead we note only the recommendation of the ax manufacturers themselves: With logs between twelve and sixteen inches in height, your chopping block should be no higher than your knees, and probably even lower.

The kind of wood used is less important, though wood that is hard and twisted is not liable to splinter as easily as other sorts, which makes elm a popular choice. What does matter is that the top and bottom of the block are sawed level and perpendicular. Good chopping blocks can be made from large trees that have been felled by machine, especially if you can persuade the operator to cut you off a wide conical section from near the base of the tree. And what about those logs that have been cut slightly crooked under difficult conditions in the forest? The sensible woodcutter deals with this problem by using a second block, one with its top sawed at an incline so that angled logs can be balanced up against the bias for chopping. That's a little trick that will keep you smiling for years.

Some people knock nails into the block and grind the heads down to a point about a half inch long to pin the log at the bottom so that it stands upright. And the real enthusiast may well stain or tar the underside of the block, to stop it from rotting upward from the base, which would eventually soften the block and steal energy from the chopping. A car tire on the top of the chopping block can greatly improve your working conditions. This holds the split logs in place so that you don't need to pick them up after every blow, and such a guard makes it a quick and simple matter to chop for kindling—all those wood chips, offcuts, and splinters that seem so unimportant at the time are exactly what you're going to need most of all on cold winter mornings. An elastic strap can also be used around big logs.

Splitting Techniques

The difference between splitting fresh wood using a good tool and—well, and almost anything else—is enormous. The drier the wood, the more tightly will the fibers adhere to each other. The cell walls are also softer when wet.

Firewood must be split, because bark retains the moisture in much the same way that the skin does in an orange. Splitting the wood greatly accelerates the drying process.

Chopping up large trees and leaving the logs to lie with the intention of splitting them later, "when I've got time," or "when I can get someone to help me," is just asking for trouble. The result will be, inevitably, moldy, worm-eaten wood, a scolding at home, and a trip to the garbage dump four years later.

If you live somewhere where the winter temperature regularly falls below the freezing point, the cold and damp together can be used to your advantage. Most fresh wood will split with a light blow in subzero temperatures. An old trick used by Norwegian woodsmen in the late winter was to smear the chopped logs with a handful of snow at each end. In the morning the sun melts the snow into the wood; come nightfall the moisture will freeze, and with luck the log will split at the first blow of your ax the following morning.

Knots and irregularly shaped cuts of wood will always pose a problem. It is an advantage to cut such wood short, because a rule of thumb says that doubling the length of the log requires four times more penetration before it will split. Logs should be cut to avoid knots at the top and bottom. A hydraulic splitter makes short work of whatever it encounters, even where the wood is long and difficult, and a helicoid wedge and sledgehammer can also be very effective. Both methods, however, are labor-intensive, and often the quickest way to dispose of difficult logs is the hydraulic splitter. For ordinary straight-grained wood, the splitting ax is often the best tool, especially when using the car-tire method of containment (see page 99). Working in a regular, rhythmic fashion, with a swing in which the upper hand slides down the shaft in a controlled glide, is your best guarantee of an efficient and safe session with an ax. Speed and precision are more important than brute strength, and skilled woodcutters often employ a technique known as the "sweep" in the last phase of their swing, dragging the ax slightly downward and inward as a way of increasing the speed. The knees should be bent just prior to impact, giving more power and more accuracy to the blow. Following the blow all the way down with the body lessens the chance of injuring yourself in the case of a miscued strike. Do this, and the ax will hit the ground in front of you; standing up straight you run the risk of an out-of-control ax hitting your ankles or knees. Splitting wood with an ax is mostly a question of technique and mental attitude. The first thing is to make sure the blow is executed quickly. It is often the case that once you have made up your mind that that log is going to split, it does split—because your blow is quick and decisive. Think of yourself as a karate master who can split a brick with his bare hand. In this technique, known as *tameshiwari*, the focus is not at a point on top of the brick but directly below, and the blow falls as though the brick were not there at all. This is one of the great liberations of splitting logs: No hesitation, no doubting—just hit it!

"Reading" the tree will tell you where to aim your blow, because wood usually follows and divides along a naturally occurring ax. In the spring these are evident as small splits that appear a day or so after felling. Wood is often easier to chop from the top down, with the log on your chopping block standing in the same direction as did the tree when growing in the wild. The edge should cross as few rings as possible, because a blow that strikes across at an angle will encounter the resistance of the natural direction of the fibers' growth. The ax should strike between two knots, or directly onto a knot. If the first blows don't lead to fracturing, try somewhere else. With large logs there is little point in bringing the ax down right in the middle—the blade will get stuck in a foaming slit of sap. Better to work around the edges and chip the wood off, like the cords of a circle. The flat bark from the outermost layer will make a good roof to lay along the top of your woodpile. It is a good plan to chop wood in various sizes, both thicker logs of the harder tree sorts, and smaller logs of, for example, aspen or spruce. As we shall see later, this will provide wood suitable for all outdoor temperatures and keep air pollution to a minimum.

Wood does not have to be split in an upright position. The old woodcutters were expert at splitting it as it lay, striking directly into the bark. For this you need a splitting ax with a narrow head, though some prefer to use a felling ax. The logs are firmly struck near the end, and then several more blows follow the split upward. To begin with, there will probably be a lot of wild misses and flying chips, but with practice this can become a very quick way of working. With a single thick log as a base, several more logs can be split before it becomes necessary to bend down again. This was once a much-used method of chopping cordwood twenty-four to thirty-two inches long out in the forest. In the really old days, before the introduction of the saw, the technique was ubiquitous. There are drawings from all over Europe showing logs being split in this way as early as the fifteenth century. Only with a saw, obviously, was one able to make the sort of level cut that meant a log could be balanced upright on a chopping block.

But the method has its uses even in our day, and is especially practicable for those splitting and drying wood with limited space at their disposal. Here, wood twice the normal length can be a good alternative, because more of the work can be done out in the forest, and because wood twenty-four to thirty-two inches long is easy to transport. The only part of the job that needs to be done at home is to cut it in two and stack it in a dry place.

The Logistics of Wood

Back in 1920 Henry Royce, the engineering half of Rolls-Royce, noted that "Each time a material is handled, something is added to its cost, but not necessarily to its value," and his observation might well be applied to wood. Because there is no getting away from it, wood is heavy. True enough, the subjective weight of one pound of wood is well below one pound, in much the same way that a pound of silver (probably) weighs less than a pound of rubbish. But the pleasures of personally handling and transporting your own source of energy do come at a price.

So for efficient and pleasurable dealing with firewood, we should heed Royce's words. A quick calculation tells a gruesome tale of just how physically demanding it is to harvest firewood. If the procedure is reduced to a practical minimum there are six stages: dragging the timber to the trailer; lifting it to the crosscut saw or the sawhorse; lifting the logs for splitting; stacking the logs; and moving them from the stack into the wood basket and, finally, to the stove. A typical Norwegian cord of mixed firewood weighs four thousand pounds (1,800 kilograms) green and two thousand pounds (900 kilograms) dry, so each cord of wood involves handling a total of 9 metric tons. From the time the tree is felled out in the wood until the last ash pan is emptied a woodcutter with an annual consumption of four cords will have handled 36 metric tons of wood. Every additional stage involving unseasoned wood adds an extra 1.8 metric tons per cord to that sum, and two thousand pounds (900 kilograms) for every additional stage involving seasoned wood.

So a thoroughgoing rationalization is vital to make sure you get real value for all your hard work: Plan each stage so that the wood has to be moved as few times as possible. Best of all is to chop and split the wood at the same place that the tree was felled, but this is rarely possible. If the land is sloping, utilize this to make sure that the wood will always fall *downward* toward the next stage of the process (see slope-felling, on page 44), or at least eliminate the need to lift it again. If using a crosscut saw, for example, place a wheelbarrow so that the logs fall directly into it. A good idea when cord stacking is to cut the wood so that the logs lie in an oblong row, and then move the chopping block along as

According to an old saying, wood will dry well if there is space enough for a mouse to run through the whole woodpile. Stack made by Liv Kristin Brenden, of Brumunddal.

you split and build the stack. Woodstove users in apartments or houses with limited outside space often use the split-and-dry method, whereby the wood is dried out in the forest and only then transported to the garage, the veranda, or the cellar.

In many cases the best drying site is situated some distance from the house. If, for whatever reason, you are unable to keep the wood in a nearby garden shed, you can solve the problem by transporting enough for the needs of a day or two from the stack to somewhere closer to the house, a sheltered spot with a small chopping block and a hand ax to chop that kindling you should have remembered to chop earlier. Another delightfully simple and useful convenience is the wood bag—simply a good-size carrying bag capable of holding thirty to forty liters. Most of those available, however, are much too frail—they tear in the cold, or the handles fall off. If there is a sewing machine in the house then four or six of these made of canvas or sailcloth and sewn together with a sturdy handle that goes all the way around the opening will prove a treasure. They take on a lovely patina and last forever.

Banana boxes are usually 14.5 inches wide and might have been made for storing kindling. Another useful tool is a dolly or wheelbarrow chassis with a box mounted on top, perfect for transporting a day's supply of logs from the woodshed to your back door.

Both hands are kept pretty well occupied doing such jobs, and you can make it easier on yourself by planning ahead—a spring-loaded door and outside lights activated by motion sensors will smooth things along and impart to the whole operation a pleasing feeling of being in charge.

ICA

ELGÅ: THE WOODSHED IN THE SOUTH WIND

Close to Femunden Water, the third-largest lake in Norway, lies Elgå, a tiny community that has always lived close to the forces of nature. The village was inaccessible by road until 1956, and the fifty people who live there enjoy harsh weather conditions, bitter cold, and fabulous natural surroundings. Domestic heating is provided by a full spectrum of woodstoves. The heart of the community is its one and only store, the well-stocked and well-run Peder Røsten. Fuel is dispensed from a battered old pump that still uses a mechanical counter.

Not far from the shop is a small, neat house with an equally neat woodshed. This is the home of Ole Haugen, born in 1926. Ole is a real firewood enthusiast and all heating in the house comes from firewood he has prepared himself.

He makes an admission, however, as he offers coffee: "If it gets cold enough, I will turn on the little electric heater in the bathroom, and on a really cold night I might cheat sometimes and leave some electric heating on in the kitchen."

On the other hand, Ole's idea of what constitutes a really cold night is not everybody's: "It never gets that cold here in Elgå. Usually not much under twenty below Celsius. Occasionally thirty. But I do recall one winter evening many years ago when we'd been to a show in Tufsingdalen, on the other side of Femunden Water, and when we came out it was minus forty-three."

Ole has been a joiner and woodcutter all his life and has always cut his own firewood. In his childhood the usual thing was to burn dry pine, felled using a *svans* (a large two-man crosscut saw) and dragged back home to the village by horse. Heavy, time-consuming, and exhausting work.

"The chain saw was a revolution when it came," says Ole. "Those first saws could weigh anything up to twenty kilograms and a lot of people didn't take to them, but to me they were a miracle to work with."

In time, he says, foresters developed a real fondness for the chain saw. Elgå and the surrounding districts became typical "Husqvarna towns"—they didn't think much of saws made by Stihl. Toward the end of his working life he switched brands and his last two saws—both of them now in his work shed and displaying all the signs of years of loving care and hard use—are Jonsereds, a 590 and a 2051.

Ole used to get all his firewood from a birch wood growing high up on a nearby hill. Because of the undulating terrain and the distance involved he

Ole is a Jonsered man. He keeps his trusty 590 and 2051 chain saws sharpened and ready for action.

usually leaf-felled the trees, leaving them unlimbed to make transportation easier. As the leaves continue to grow the moisture is drawn up out of the trunk, and after a few weeks the tree can be limbed and sawed up in the usual way. Because the logs are now much lighter, the method is especially suitable for places where the wood has to be transported over a distance.

"In the spring I used to go up there regularly to check on the leaves. The rule was to fell the tree once the buds were the size of a mouse's ear. Up in these parts that would generally be about the start of June. You couldn't wait about, because around that time of the year the sap rises fast, and once the leaves are in full bud there's a lot of moisture in the tree."

With the trees felled, the timetable for the rest of the work is less critical. So if you're a little behind with the chopping and splitting and final drying, there is no great harm done. Woodcutting using the leaf-felling method also makes it easier to fit in around the other jobs.

Yet Ole tells me, "If you have a wood that's close by, and plenty of time in the spring, then I still think the traditional felling in late winter is best. The wood

The drying shed has wall dividers that effectively turn it into four compartments. Doors make up the whole of the front wall, and the rear wall can be removed, giving access from both sides. The design ensures good air circulation and well-dried firewood, and simplifies access to the logs.

is easy to handle and you get good use of what we call the drying months—April, May, and June."

Over the years Ole has observed how important it is to stack the wood so that the wind comes at it from the most advantageous angle.

"Wind blowing through the stack is what gets it really dry. Heat is important, but damp wood should be out in the sun and wind as much as possible."

Like other firewood enthusiasts, Ole likes to try out new things and refine his understanding of the factors that influence the quality of the firewood. During his life he has experimented with a number of different designs for his woodshed. Now, just past his eightieth birthday, he has finally come up with one he thinks is satisfactory.

This shed is long and narrow, with a rear wall that can be removed. The four doors at the front run the full width of the shed, and when these are opened the wind blows through the entire stack. The shed is positioned so that the south wind strikes the open rear wall. Inside, the logs are stacked so that the wind blows along them lengthwise. The dark corrugated iron roof retains heat, which speeds up the drying. The removable rear wall means that wood can be stacked from the front or the back. As winter approaches the doors are closed or the back wall placed in position so that snow does not blow in. The design makes maximum use of all available space, and the wood is easy to get at. And, of course, the shed has its own permanent reserve for the extra-cold winters.

"Around these parts hardly anyone keeps their wood in unstacked piles. Nearly everyone does their own firewood, and they'll stack it. Some of the modern woodcutting machines store the wood in great sacks on a pallet, but that doesn't really count. You get good wood from it, but it's not been done by hand."

Ole is the kind of man who would rather sing the praises of others than his own. His stacks, he says, are simply practical constructions that do the job. But seventy years of experience tell their own story. The ends of his stack are so neat it looks as though the whole thing had been trimmed on both sides with a huge circular saw. Not a single log has been laid crosswise. Even twisted logs have found their places in the stack, without compromising the stability of the whole.

"My method is very simple. I do my chopping, splitting, and stacking in small doses. That way I don't get too stiff, and the wood doesn't lie too long on the ground. The secret of an even woodpile is to learn the trick of knowing what sizes you need to make a stable structure. It's easier to find the right logs and cut them when you're not trying to do the whole job at once. And I also allow for the fact that the wood is going to shrink a little as it dries, so I build in a slight inward lean against a support, so the stack won't topple forward that easily."

With advancing age Ole has had to adapt his working methods. When he began feeling too old for the chain saw he started having his wood delivered as logs. But he still did the cutting, splitting, and stacking himself. Later, after a stay in the hospital, he was specifically warned not to chop wood because the force of the blow transmitted itself directly to the skull. "Time to put that ax away," the doctor told him.

Ole had the answer to that. Not long after returning home he went out one day and bought himself a hydraulic splitter.

CHAPTER 5

THE WOODPILE

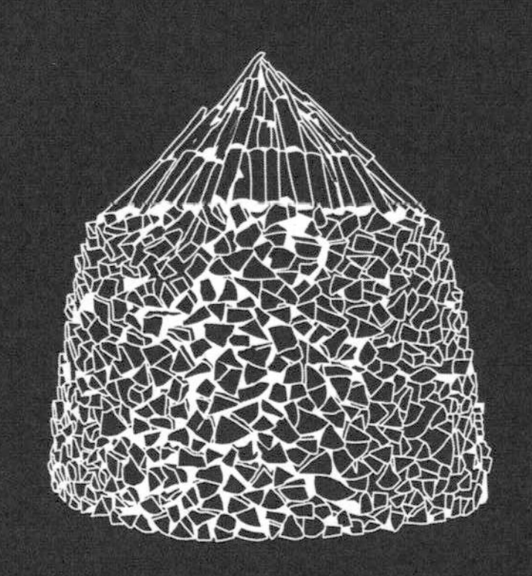

It took a while, but that didn't bother them, as long as it turned out the way they wanted.

—Nilas Tuolja, a Sami, speaking about Sami who had grown too old for any work other than stacking dried spruce

You know exactly where you are with a woodpile. Its share price doesn't fall on the stock market. It won't rust. It won't sue for divorce. It just stands there and does one thing: It waits for winter. An investment account reminding you of all the hard work you've put into it. On bitterly cold January mornings it will bring back memories of those spring days when you sawed, split, and stacked as you worked to insure yourself against the cold. There's that twisted knot that just wouldn't surrender to your ax. And isn't that the log you pushed in at the wrong angle, making the whole pile collapse? Yes, that's the one all right. Well, winter's here, and now it's your turn to feed the flames.

And on the subject of woodpiles, let's hear from Thoreau once more. "To affect the quality of the day, that is the highest of the arts" was his mantra, but his most famous words on the subject of wood are probably these: "Every man looks at his wood-pile with a kind of affection. I love to have mine before my window, and the more chips [around the chopping block] the better to remind me of my pleasing work."

Here is the majestic result of all your hard work. And the sight of a woodpile is the sight of security itself. A lot of people like to build it on a spot where it can be seen from the kitchen window. It makes a marvelous view. Like a geological layer on a mountainside, the pile is a reminder of the work you did last year, with the colors of the rarer sorts of wood showing up in lateral stripes, and unsplit wood from small trees tracing small, circular patterns within the expanse of wood chopped from the more common trees the previous year.

Good to look at a woodpile may well be, but its primary purpose is the essentially practical one of ensuring that the wood dries, and remains as dry as possible. In the final analysis it is the stacking of wood that dictates its quality, in regard both to moisture content and to appearance. For this reason many enthusiasts ready their wood in two distinct stages, first drying it in the open air and then, sometime in the autumn, moving it to the woodshed. Whichever it is, the lapse of time between the standing tree and the chopped and split logs in the woodpile should be as short as possible.

Wood is best (meaning driest and least exposed to fungi) when dried *quickly*. Wood that has been dried quickly also seems to absorb less moisture in the autumn. Actually, a great deal depends on the first month. The methods employed to dry timber intended for use as building material and in furniture making often focus on the need to *prevent* the wood from developing cracks and so are of little interest to the woodcutter. As much surface area as possible should be exposed to wind and sunlight. This is best achieved by positioning the pile so that the wind blows freely through it from all angles, ensuring at the same time that it is sheltered from the rain. Just letting the logs lie on the ground and hoping for the best will not do. Compare it with how you deal with the washing: You would never put it out to dry rolled up in a ball on the ground. If there is a choice between storing the wood somewhere warm or somewhere well ventilated, the latter is the preferred option. It is wind that gets wood really dry.

You hear rumors of wood so densely stacked that you could hardly slip a cigarette paper between the logs, but such stacking should be done only with wood that is already dry. In fact the ideal way to dry wood is to pile it as loosely as possible short of collapse. The old-timers had a rule for this too: Unseasoned wood should be stacked so loosely that a mouse could run through the tunnels. If wind and sun are allowed to do their work, it will dry quickly. In the late

PAGE 110 A low square pile is a good choice for short wood.

PREVIOUS LEFT At the Kuremäe monastery in Estonia, the nuns build these round stacks, which are six and a half feet (two meters) tall. The wood is twenty-four inches (sixty centimeters) long, and they use ladders for most of the process.

PREVIOUS RIGHT Erling Gjøstøl, from Ådalsbruk in Norway, with the result of his springtime work. He built a standing carousel (which you can see in the background), a method also used by the Sami people in northern Norway.

summer it can be transferred to a shed that will provide shelter from the rain and snow, and there you can stack it as densely as you like.

Pallets or poles laid in parallel along the ground will protect the stacked wood from moisture rising from below, but protection from rain is also needed. Black sheet iron is ideal for the purpose. The sheet retains the heat, helping to speed up the drying process, and can be positioned so that the upper layers of the wood are aired. Plastic or tarpaulin can also be used, though these will never allow as much air to circulate as a hard roof. Under no circumstances should the wood be wrapped, because the air within gets so damp that the wood is exposed to attack by mold and fungus and will not dry. Many firewood enthusiasts leave their wood without a top covering in spring, because there is more moisture that needs to get out than the small amount of water that might make its way in during the spring after a few showers. Some people actually expose their split logs to rain or water for a short period at the very outset of the drying process as a way of washing out the sap or the tannin in oak, or to get the wood to swell. The logs are then dried normally and develop large splits. Generally speaking, however, the risk of fungus is so great that this procedure is not recommended save for those who enjoy trying out new things.

It is possible to stack unseasoned wood in a woodshed, but provision must be made for air to circulate easily around it. A large pile of wood in a shed that is not well ventilated will take an age to dry and is more susceptible to mold and fungus.

Avoid anything that might seem like a *temporary solution*. Typically this might be a tentlike structure made of plastic or a tarpaulin. These invariably collapse and your wood will end up moldy and rotten.

The basic requirement for a loose construction and a well-ventilated site need be no hindrance to the aesthetics of your woodpile. And it will be a practical aestheticism, because a well-assembled stack makes for dry wood that in turn makes for good heat in your stove. It will also be able to withstand the buffeting of the winds. Hard work and imagination are called for, so results can reveal something about the builder's own character. But there are dangers: Good sculptural ideas can turn out to be trickier than you had expected, twisted wood creates instabilities, and shrinkage has to be taken into account. From unseasoned to bone-dry, the volume of the stack will shrink by between 7 and 20 percent (the exact figure depends on the kind of tree), and with the passing of spring this can cause it to sag and collapse. Wood piled in the winter can give you a nasty surprise when spring comes along and one fine day you discover your beautiful woodpile collapsed in a jumbled heap. Suspecting the neighbor of sabotage usually turns out to be a dead end; the more likely explanation is that the ground frost has melted, causing the foundations to move.

Stacking is an aesthetic and a practical challenge, so much so that in the late nineteenth century, in the heavily forested state of Maine, young American women considering a potential husband were advised first to consult a piece of folksy wisdom that revealed the young man's character based on the way he stacked his wood. In all Scandinavia it is also common wisdom that you can tell a lot about a person from his woodpile. For those looking to marry, the following list may be used as a rule of thumb.

Upright and solid pile: Upright and solid man
Low pile: Cautious man, could be shy or weak
Tall pile: Big ambitions, but watch out for sagging and collapse
Unusual shape: Freethinking, open spirit, again, the construction may be weak
Flamboyant pile, widely visible: Extroverted, but possibly a bluffer
A lot of wood: A man of foresight, loyal
Not much wood: A life lived from hand to mouth
Logs from big trees: Has a big appetite for life, but can be rash and extravagant
Pedantic pile: Perfectionist; may be introverted
Collapsed pile: Weak will, poor judge of priorities
Unfinished pile, some logs lying on the ground: Unstable, lazy, prone to drunkenness
Everything in a pile on the ground: Ignorance, decadence, laziness, drunkenness, possibly all of these
Old and new wood piled together: Be suspicious: might be stolen wood added to his own
Large and small logs piled together: Frugal. Kindling sneaked in among the logs suggests a considerate man
Rough, gnarled logs, hard to chop: Persistent and strong willed, or else bowed down by his burdens
No woodpile: No husband

The Small Arts of the Woodpile

The first rule is to ensure that the pile you build is appropriate for the kind of wood you are stacking. Twisted wood is best stored in low piles, drying bins, or at the center of circular stacks, but there is no limit to what you can do with straight logs. And the longer they are, the easier they are to stack. Short firewood under 10 inches in length will collapse as soon as you look at it and should be stored in a sack or piled up against a wall. If your stove can take it, logs between 14 and 16 inches long are a lot easier to build up into a freestanding woodpile, whereas 24-inch cordwood will give you a stable construction regardless of height, shape, and wind conditions.

The most stable constructions are achieved using logs that are roughly the same length. Learning to build a pile may take some time, and it is not a job that can be rushed. Each log needs to be thought of as a brick. You will need to become familiar with the factors governing the inner stability of the stack, see the way quartered and halved logs slot together, and continually check the pile for instability and unevenness. Any unevenness will be reproduced as the pile rises, and special account will have to be taken of logs with knots and other irregularities by, for example, placing logs with similar irregularities upside down on top of them. Logs that have been cut in half are the most stable, and these are particularly recommended for use in a crisscross pattern at the ends of the pile.

Logs dry best when the surface contact between them is minimal, but as the stack moves while drying, the logs will tend to sink down, with the attendant danger of the construction listing. To give an example, quartered logs that balance each other with their ends pointing downward can prove very unpredictable and it is advisable to pile these so that they rest on each other on two surfaces. Many people use the so-called crossbar, a log stacked sideways on the rest of the pile, as a way of adjusting for a potential irregularity. The practice is frowned upon by aesthetic pedants partly because it disturbs the lines of the stack, but mostly because it reveals the presence of an unstable section lower down in the pile.

In Norway, discussions on the vexed question of whether logs should be stacked with the bark facing up or down have marred many a christening and spoiled many a wedding when wood enthusiasts are among the guests. In the autumn of 1998 the correspondence columns of *Østlendingen*, a local newspaper in Hedmark County, hosted a heated debate on the subject. The bark-down faction maintained that bark turned upward acts as a roof that hinders the escape of moisture from the logs. A scientific research project (which included CAT scanning in a Norwegian hospital and the assistance of a local secondary school!) demonstrated that though there is some truth in this, the overall effect is negligible. Yet there are two distinct schools of thought on the matter here in Norway: bark up, along the coast; bark down, inland. The reason is that coastal rain is often hard and lashing, and because bark is water-repellent, the wood stays drier with the bark on top. Inland regions don't have the same problem, provided the wood is well protected above. Bark down, however, is a standard rule for the bottom layers as a protection against ground moisture, and bark up on piles that have no top cover.

Today the art of stacking wood in Norway is confined to a few distinct

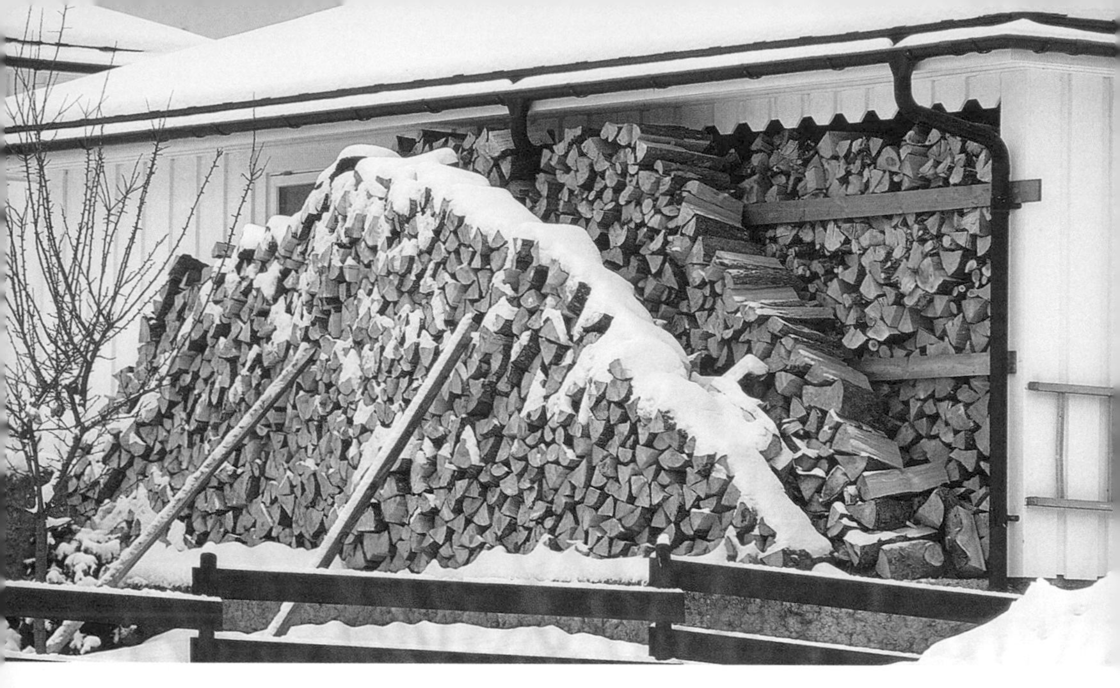

A sun-wall stack in three sections, reinforced for the winter.

styles. Piles built against walls offer few possibilities for variation and the circulation of air is not optimal, but the construction is stable. A drive through rural Norway will soon show how widespread this method is. In Finland the situation is different; there, fire-safety regulations prohibit the stacking of large quantities of wood against an external wall. There are also regulations governing how much wood one may store within a specified distance from a dwelling place, with the result that the country has developed a strong tradition of freestanding piles. Such piles are more difficult to build, but they can be sited where wind conditions for drying are advantageous. A good choice is where the drying rack stands, or—more realistically—the second-best place, after the place where the drying rack stands.

The Norwegian Sun-Wall Woodpile

This is a real classic—where regulations permit it. It is functional, stable, and easy to build. The wood is piled against the wall of a house, preferably a sun-facing wall and—for best air circulation—not too close to other buildings or trees. Pallets or long poles are used to keep it off the ground. If the logs are small it is a good idea to attach planks, four or five every square meter, at an angle of ninety degrees to the wall. When the logs press up against these planks the pile locks and stands firm. It is also sensible to leave a small passage between the wood

One of the many sculpture stacks that pop up in rural parts of Norway during the spring. This one is made by Bjarne Granli at Drevsjø.

and the wall to facilitate the flow of air. The higher the structure the greater the risk of collapse, so better to have a pile tilting inward than outward. Often, however, the best way to avoid the risk of collapse is just to nail up a retaining plank that runs the length of the pile, and this may be a requirement if the pile is up around the height of the gutter. With the first pile in place a second can be built against it. The inner pile gets less air, but if you have enough wood to last you a couple of winters this is a very useful method. A three-layer wall, a variation on this style, is shown on page 118.

The Firewood Wall or Firewood House

This is a variation on the sun-wall woodpile. The external wall is covered to its full height in wood, completely obscuring the original surface. The effect is particularly striking if all four walls (excluding doors and windows naturally; one mustn't get carried away!) are clad in firewood in this fashion. The result is reminiscent of an even more revolutionary construction, the freestanding woodshed, which is built of wood in its entirety. This is defined as a stack and so is not covered by, for example, the Finnish fire-safety regulations. A woodshed with a surface area of about fifteen square meters will require about eleven cubic meters of wood, and the wood should be between sixteen and twenty-four inches long.

Round stacking step by step. The horizontal circular base is built up into a cylinder, with the logs in the middle providing support and preventing inward collapse. When the structure is about 3.5 feet high the logs are laid in toward the center, and the whole thing is topped off with a layer of small logs with the bark facing upward. Stack made by Ruben Knutsen, of Hamar.

The Round Stack

Known as the beehive or the Holzhausen, the round stack is an outstanding form of woodpile once widely used in Norway, but now almost obsolete. It is not easy to make, and if it starts to collapse the whole thing goes. But a successful round pile has much to recommend it. It makes good use of the available space and can accommodate twisted wood, and, if it's properly constructed, rainwater will run off the outside so it does not need a top covering.

The basic idea is to lay the logs in a large circle. If you are willing to take a stab at building a round pile with a really large diameter then one or two smaller inner rings are possible. Logs that are short, twisted, or for some other reason difficult to stack go in the center. The pile is raised horizontally to a height of about 3.5 feet. At each subsequent course logs are laid slightly in toward the center, with the center being filled up as you go along to ensure stability. The whole is topped off with a roof of flat logs, laid with the bark upward to keep the rain off.

Some put a pole at the center and mark it to show the height of the pile. This enables them to see how much the wood shrinks, and from that to work

Logs 24 inches long, also known as cordwood. After drying, the logs are cut again and moved to the woodshed. Stack made by Liv Kristin Brenden, of Brumunddal.

out how dry it is. Others put a stovepipe in the middle, removing it afterward as a way of improving the circulation of air. Round piling is described in detail in the next section.

Cord Stacking

This is a good choice, being aesthetically pleasing and easy to build, and, not surprisingly, a long-established form because it affords outstanding drying conditions. These stacks are easiest to build with long wood—traditionally sixteen or twenty-four inches—layered on poles. Starting the stack with a tower of logs laid crisscross will give you stability. An alternative is to knock supporting posts into the ground.

The best drying is achieved with a pile laid east to west; this allows the south wind to blow straight through when it strikes the pile on its long side. Such stacking also makes it simple to measure out the classic Norwegian cord of wood. For wood about 2 feet long, a section of the pile 3.5 feet high and 13 feet long will give a cord. The traditional North American cord is 4 feet high, 8 feet long, and 4 feet wide. Any stacking that produces a corresponding volume

(128 cubic feet) is equivalent to a North American cord. It should be covered with fairly broad wooden plates so that the rain runs off the back of the stack. An alternative is to create a high point in the middle and use long roofing plates that slope down toward the two ends, or else you can build the stack on sloping ground. Wood that is particularly long (between two and three feet) can be cut by positioning the chain saw over the middle of the stack and cutting downward, but the real purpose of wood this length is to make it easier to transport from the forest to the house. The technique of cord stacking is described in more detail on pages 176–179.

The Closed Square Pile

This sturdy and space-saving method is a good one to use with short wood. Closed square piles are almost always raised on pallets arranged in a large square. The wood is stacked in several rows, and in the case of really big squares you will have to start in the middle and work outward until you have filled the whole area. Crossed logs at the corners give stability to the pile. This method makes it possible to stack wood of different lengths in the same nice-looking pile, but the larger the stack the more preparatory work and checking are required (see illustrations on pages 19 and 182–183). The flatness of the structure makes a corrugated-iron roof advisable to keep off the rain. A tarpaulin should be avoided for large square piles, as it is likely to cause dampness in the air around the top of the pile.

The Open Square Pile

The open square pile is outwardly similar to the closed square pile, with the same crisscrossing corners, but within the outer wall the wood lies in an irregular heap, much as in the round stack. This makes it particularly well-suited for crooked and uneven logs, those, for example, from large trees with thick branches. Unlike the closed square pile, the open square stack is built from the bottom upward, creating the full width from the beginning (see illustration on page 80).

The Standing Round Stack

This is the round stack's vertical cousin, requiring a little more equipment but offering unique advantages. The stacks are completely circular and are held together with load straps or wire. You need a round frame, preferably made from two semicircular sections of strong iron. These are fastened together with bolts at the top and bottom. The ring is mounted on two logs and filled with wood, then the strap or wire is tightened around the logs. The ring is then split and the two halves moved to the place where you intend to place your next stack. Finished round stacks can be rolled or lifted with a crane onto a trailer. A corrugated-iron roof can easily be fastened across the top. The largest practical diameter is about 5 feet, and a stack this size should consist of logs 1.5 to 2 feet long so that they don't slide out (see illustration on page 87).

The Sculptural Stack

This is a generic term for all woodpiles that exhibit a distinct structural pattern or play of colors, be it abstract or figurative. These are popular in Norway and each year local newspapers run photo competitions to pick the best of them, which are immediately "liked" by several hundred on Facebook. Labyrinths, sculptures, and abstract shapes are equally popular.

In 2012 a retired electrical engineer named Ole Kristian Kjelling, who lives in one of the country's most sparsely populated areas, made newspaper headlines throughout Norway when he created a sculpted-woodpile portrait of Queen Sonja and King Harald V, on the occasion of the king's seventy-fifth birthday. Kjelling had previously portrayed the composer Rossini and the local mayor.

An important choice facing the creator of these sculptural stacks is whether to permit the use of artificial coloring to bring out the contrasts, or to use the natural play of color in the different kinds of wood to create the design. The open ends of oak, for example, are a deep dark brown, pine and spruce take on a fine yellowish tone when the sun is on them, whereas linden, maple, and aspen will retain their almost white color even after the summer. The speckled alder is unique and particularly charming. At the start of the seasoning process it exhibits a fabulous, almost luminous reddish orange. If the logs are stacked upright with the bark facing outward a new spectrum of colors and textures becomes available, from the white of the birch bark, to the brown tones of the conifers, to the gray shading of the broad-leafed trees.

The V-Shaped Stack

This stacking style was used in the old days to start the drying process while still out in the forest. It was also the unit used to reckon the woodcutter's piecework. The stack consists of long, unsplit logs laid in a V shape in such a way that the ends cross each other. The rule was that the logs were to be 3 meters (10 feet) in length, and the height of the structure one meter (3.3 feet) at the cross. The logs were to be laid at an angle of ninety degrees to each other, with the thickest

end outward, and strip-barked along two or three sides. The logs at the bottom rested on stones or stumps.

The 3-meter rule was related to a now obsolete measure known as the greater cord, this consisting of unsplit and unchipped logs with a volume of 2 x 2 x 3 meters. The ubiquitous use of the meter measure was also one reason why the forest manager's walking stick was always exactly one meter long.

The V-shaped stack is little used today, because for most people it is now a fairly simple matter to transport the wood so that it can be cut and split at once. Along a stretch of the E6 highway south of Hamar, in Norway's Hedmark County, landscape architects have built several decorative V-shaped piles in honor of this old forestry tradition.

The Standing Carousel (Sami Carousel)

On mountain heaths and in other dry areas, cut whole trees, if thin enough, will dry well if limbed, strip-barked, and left to stand upright like a *lávvu* or tepee. In Finnmark, the most northerly county in Norway, this method is widely used, especially by the indigenous Sami. Practically the only trees that grow on the Finnmark moors are thin, twisted birch, and the air is so dry that the poles will dry sufficiently well provided they are partly strip-barked. Come winter the wood is transported to the houses by snowmobile, cut, and tossed into an outbuilding. Such wood, being largely twisted, is rarely stacked.

The Drying Cage

Not technically a woodpile, the drying cage is a quick and handy drying method—simply a large, barred crate with plenty of space between the bars into which the logs are tossed. It is a good choice for the pragmatist with small children and not much spare time, though it does not score highly on aesthetics. But the drying cage is an excellent supplement to more fanciful structures. Size is no object, and it will not reject offcuts, roots, or twisted wood. Drying conditions are quite good since the logs are all in an open heap, though the cage should not be too large. One can easily be constructed from the kind of mesh reinforcement used in molding concrete floors, which is sold by building suppliers. The gauge is so heavy that the cage retains its shape even when full, and it can withstand the impact of heavy logs. Use pallets for the foundations. The mesh should be wrapped around them and secured with cramp irons and wire. If the cage walls are high, use a wooden frame for support. Leave an opening so that you can get at the wood, either by threading planks through the holes in the mesh or by making openings in the mesh that can be hinged back with wire. Corrugated iron will do fine for the roof. Since the lower parts of the cage are exposed to heavy rain and drifting snow, these should be protected with, for example, plywood before the winter comes. When wood is removed, a small mattock or similar tool is handy for dragging logs toward the opening (drying bins are illustrated on pages 17 and 81).

The Woodshed

Few joys can match that of fetching wood from an old woodshed with a well-worn floor, chipped and scarred up and down its walls, stained with resin and other signs of long and honorable use. But as we noted before, logs should be thoroughly dry before being moved into the shed. Once there it makes no difference how tightly they are packed, and the pedants among us can give full rein to their love of precision by tapping the ends afterward with a ball hammer to get them flush. It is a good idea to use supporting pillars on the inside to stabilize the stack, and try to arrange things so that you have access both to finely chopped wood of the lighter sort and to large beech or oak logs, for example, for when the really cold weather comes. As the damp days of autumn begin, the wood inevitably absorbs some moisture, so it is important to ensure the floor is dry and that the walls can breathe, that you have ventilation columns at the top and bottom. A woodshed should ideally have a dark roof that will increase the temperature inside in the summer. The construction of the roof should be such that condensation does not form when the weather changes, otherwise the roof will drip irritatingly onto the wood. A good rotation is to fill the summer playhouse with your winter wood.

The Open Woodshed

A variation on the woodshed, this eliminates the need to move and restack the wood once it is dry. The open woodshed is a permanent fixture, but one or more of the walls can be removed to facilitate ventilation and give the wood the benefit of the sun. In summer the long side should be open and facing in the direction of the prevailing wind. As the threat of snow approaches, the doors and walls should be put back on. This is an outstanding compromise solution that affords even drying conditions throughout the year. Separate compartments make it possible to dry wood that has been chopped throughout the year. With such a construction it is also possible to use heavy-gauge reinforced wire mesh to make one or more permanent drying cages with hinged openings. Another possibility is to install the kind of support planks mentioned in the description of the sun-wall stack.

HAMAR: SCULPTURE IN THE GARDEN

Every spring a beautiful piece of sculpture makes its appearance outside a pleasant-looking, white-painted wooden house in the center of Hamar. Every autumn it is pulled down—or it collapses of its own accord. Evanescence is an important element of Ruben Knutsen's annual garden installation.

Ruben is a painter and sculptor. He hails originally from Arendal, on the coast, but was bitten by the woodpile bug after moving to a new home in the interior of Norway. He wanted to make something special out of the job of piling wood, so every year he wrestles into shape a new version of the classic round pile.

Woodpiles such as these were once a common sight in Norway, but the art of constructing them has all but disappeared. Ruben was inspired by a visit ten years ago to Skansen, the Swedish open-air museum. Each year he tries to improve his skills at creating the form.

"One of the first things I realized was that the diameter of the ring has to be quite large. That's necessary for the stability. Too small, and the gaps mean that the next course of logs is liable to fall into the gaps below them. Usually I go for a ring that measures about 2.2 meters across, but one bigger than that works just as well."

He uses short planks for the foundation. When laying the first three or four courses he is careful to make his "wall" as steady as possible. Here only the straightest and most evenly split logs will do.

A basic principle of the round pile is to fill up the empty middle with logs that are misshapen, crooked, or short. These are piled loose, but leaning up against the inner wall of the ring and providing support against an inward collapse. If most of the wood is of equal length and straight then there is no problem in building a pile with two rings, though Ruben says that stacking in the round is much easier if the logs are long.

"Thirty-centimeter logs aren't very stable. Forty is a lot easier," he says.

In addition to making sure the stack is perfectly round, one must check it regularly to ensure its cylindrical shape. While working Ruben studies the pile

PAGES 124–25 Each spring Ruben Knutsen works to perfect the art of the ring pile, and each winter he burns the whole thing up.

OPPOSITE Ruben enjoys having a view of the woodpile. As it happens, the pile collapsed the day after this picture was taken, but the wood was already dry and could be moved into the woodshed.

from different angles, knocking logs into place and testing for stability. It should be possible to sway the pile slightly, but not too much. Afterward it should revert to its former position. And there should be no tilt in it at all. Once a round pile has started to collapse it has to be pulled down all the way to the bottom ring. Making repairs isn't an option.

"That's something I found out the hard way," Ruben laughs. "Now I always stack the logs fairly tightly, but even so I've never yet had any problem with wood not being dry. All the wood I've stacked in the spring has been ready for use the following winter, and there's no green-wood hissing. It's because the sun and wind can get at it from all angles. Mind you, really big piles need to be that much more porous so that the innermost layer dries out."

As the ring gets higher any unevenness becomes more and more prominent and the stacking more and more difficult, so it is crucial to stabilize it all the while, inserting smallish logs to keep the ring shape level. Sometimes you have to cheat and add a few crosswise logs. The builder needs to have a feel for every single element of the construction each step of the way.

When the pile reaches about 3.5 feet Ruben begins to round it off. The ring grows smaller and smaller with each rising course, so that it is egg-shaped by the time the pile reaches 6.5 feet. He tops this off with "roofing tiles"—logs that are fairly flat and laid in an overlapping pattern so that the rain runs off them. Naturally these "tiles" are laid bark upward. Deciduous trees are ideal because their bark is almost waterproof.

To ensure a good airing, the pile is left uncovered throughout the spring and summer. During any periods of rain Ruben throws a tarpaulin over the top. In the late autumn the whole pile is pulled down and the logs transferred to the woodshed.

Ruben claims that the ring pile is not especially difficult to build, provided you take the time to make regular checks.

"I remember once, I noticed it was beginning to creep a bit, but I just ignored it, I didn't really want to know. The day before we were setting off for our summer holidays I heard a great crash and the whole pile came down. I just left it there," he laughs.

Ruben has some reservations about the load of wood that has been delivered this year. As usual it is pine or spruce, but this year the logs are splintered and crude after their encounter with a wood processor.

"There are a lot of chips and rubbish, and the bark is flaking off. It doesn't look too pretty, but I'll make sure the worst of it is hidden in the middle. That's one of the most unusual advantages of the ring; there are few other ways of making a woodpile that gives you this free space in the middle. I like the 'odd men out' among the logs, the ones that are twisted and difficult. They burn just as well as the others, and I get just the same pleasure from putting them in the stove."

The only tools he uses are a chopping block and an ax. The ax is a treasured possession, a hand-forged *stor klyvyxa* from the Gränsfors factory in Sweden. It is a little more expensive than the mass-produced variety, but it has the balance and personality of a handmade tool.

Ruben says that people often stop for a chat while he's working on the piles, and he is particularly pleased to see that one or two of his neighbors have also started making ring piles.

"I like to work on the stacking on my own. My family sometimes calls me a 'wood nut' when I'm standing out there in the half dark and can't tear myself away from the job. But when it comes to moving it all over into the shed, then I'm glad of a helping hand. And to be honest I like it when my family comments on the wood. It's good to sit at the dinner table and look through the window and see how the pile has grown."

He thinks you have to be a certain age before you can understand this feeling. Being a provider is another part of it.

"I've had a feeling for wood all my life, but when I was younger it was more the fire itself and the heat that mattered to me. As kids we used to make toy houses out of wood, but the logs were just building blocks to us. It wasn't until I started a family and got my own place that I began to think differently about wood. But I never try to pry too deeply into where the real source of the pleasure might be, and maybe that's the secret of it."

Ruben spends about four or five days on his stacking. As the summer advances he becomes prey to small anxieties: Will it collapse? It is perfectly normal for it to subside slightly, and on hot days particularly it will resonate with cracks as the logs shrink and settle into place.

"It's a part of the great yearly cycle of things. And wood lets you feel as though you're part of it all. A log that's heavy and damp puts you in direct touch with spring. It's a harbinger, a kind of preparation, a hint of something good in the winter that lies ahead. But there's nothing sentimental about it. I never shed any tears about burning the perfect log. It's not such a bad idea to have this kind of relationship with something that is as fleeting as life itself."

CHAPTER 6

THE SEASONING

The idea that dry wood was not something his father had thought of ahead of time frightened Roy.

—David Vann, *Legend of a Suicide,* 2008

In the same way a tree slowly takes up water, it will now, just as slowly, release it. Moisture is fire's enemy, and the conditions during seasoning determine the quality of firewood more than anything else. The stack is the woodcutter's delight in summer—when the moisture is evaporating in the heat, the logs crack open and the stack creaks as it shrinks.

Drying is a small science of its own, factoring in the tree species, the stack, the latitude, and the weather. Like the fermentation of beer, the seasoning of wood should be a slow and undisturbed natural process, untouched by the bustle of life elsewhere.

The time it takes is the time it takes.

As Dry as Possible

The coming of winter will show whether the wood has been properly prepared. Cutting, splitting, and stacking can have a real impact on the quality of the firewood, and the woodcutter's aim must be to make sure that it is as dry as possible.

One of the most enduring myths about wood is, as mentioned, that it should not be "too dry," or that it has an ideal moisture content. Not true. The persistence of the notion that bone-dry wood may fail to give off proper heat is probably the result of a combustion problem that can arise when burning intensively with spruce and other softer types of firewood. The condition is described on page 174, along with an explanation for the widespread belief that only wood from certain kinds of trees "burns hot."

It should be noted, though, that a few stove manufacturers state that firewood in their stoves should have a moisture content of at least 10–12 percent, probably as a safety precaution, and to prevent the combustion problem mentioned on page 174. Modern Norwegian research, using clean-burning stoves, has not proven any drawbacks, even with artificially dried firewood having no moisture at all. At any rate, in most climates firewood will never dip below a 12 percent moisture content when stored outdoors, so the principle of "as dry as possible" is the one to heed.

Insufficiently dry, and firewood will be difficult to burn, give far less heat, cause local pollution, and glaze the stovepipe—creating a greatly increased risk of chimney fires. Its main characteristic besides a reluctance to burn and a hissing sound—made by water inside boiling and whistling as it passes out through the ends of the log—is that it gives off dark smoke.

Fumata Nera

This phenomenon lies at the heart of one of the oldest and most distinctive signal codes used in Western culture. When a pope dies the College of Cardinals meets in a conclave behind locked doors in the Sistine Chapel to choose his successor. The cardinals announce the results of the voting to the rest of the world by burning the ballots in a small stove in the chapel. If the results are inconclusive the papers are burned with wet straw, causing black smoke—*fumata nera*—to emerge from the chapel's chimney. A huge crowd gathers in Saint Peter's Square to await the decision, reacting with mounting unease to each new release of black smoke as a visible symbol of the conclave's own frustration at its failure to reach an agreement. An army of television channels keeps the telephoto lenses of their cameras focused on the top of the narrow chimney.

PAGE 132 Most wood will crack during the most intense drying period, but when the log is dry throughout the cracks will close again.

PREVIOUS LEFT Perfect firewood—hard, dry, and clean. Fast seasoning with a lot of sun and wind is the best guarantee against mold and fungus. These organisms thrive when the wood is damp (especially between 25 and 35 percent moisture).

PREVIOUS RIGHT The V-shaped stack was very common in former times—it was even a unit of payment for woodsmen, and it was called a greater cord.

Sooner or later (the longest period of *fumata nera* lasted three years, from 1268 to 1271) the cardinals will make their choice, and when that happens, the ballots are burned along with dry straw. White smoke—*fumata bianca*—rises from the chimney, and the Roman Catholic Church has its new pope. Not even the Vatican, however, is fully able to control conditions inside the stove, and to avoid confusion it has in recent times adopted the practice of using chemical additives to better distinguish *fumata bianca* from *fumata nera*. Even so, as recently as 2005 there was head-scratching among the expectant masses at the rise of smoke that was an ambiguous shade of gray.

Measuring Moisture

Fumata bianca is the aim of every experienced Norwegian woodburner, no matter his religion. But before we look in detail at the matter of the moisture-content percentage of wood, we need to see how this figure is arrived at, because the measurement for Norwegian firewood differs from that used for building materials. The figure for wood shows what percentage of the *total* weight of the log is moisture. A log weighing two pounds (one kilogram) with a moisture content of 20 percent dryness contains two hundred grams of water. What is being measured is the *proportion* of moisture in the weight of the log at the time the log is weighed. This explains why a log containing 20 percent moisture will be *half* the weight of the same log containing 60 percent moisture. As the moisture decreases, the total weight that we are calculating from also goes down.

All moisture will reduce the firewood's efficiency and result in less heat. A minimum of 20 percent is required if wood sold in Norwegian shops is to be described as "dry"; but there is no problem in getting wood drier than that, certainly down to 15 percent, if felling has taken place at the right time and the drying conditions have been good. Less than 12–14 percent is rare for wood dried in the open, and even old and yellowing panels in a living-room wall will contain 7–8 percent moisture in the driest winter. When it comes to performance in the stove, the difference between partly seasoned wood (say, 25 percent moisture content) and really dry wood (about 17–18 percent) is surprising. The dry wood will provide considerably more heat and be much easier to light than one might suppose from the figures alone.

Drying Time

In freshly cut wood, moisture can make up as much as half the weight (sometimes more), so even a medium-size woodpile contains several hundred liters of moisture. A Norwegian cord (2.3 m^3) of birch felled in winter must release between seven hundred and eight hundred liters of water before it is ready for the stove.

The good news is that wood dries remarkably fast if it is chopped and split soon after felling. Many people maintain that *all* firewood needs at least one year and preferably two to dry properly, but in most cases this is either a country superstition or the result of experiences with wood that has not been chopped into logs straight after felling, has been felled too late in the year, or has been stacked in a poorly ventilated place. In perfect conditions, short logs (about twelve inches) can be ready for burning after just two months, if the drying is done in the spring and the circulation of air around the stacked wood is good. In climates where the atmosphere is moderately damp, most types of wood, including the hardwoods, will be ready for use by the winter if they are stacked in spring.

There are exceptions, of course. Oak usually needs two seasons to dry properly, and in very humid climates two years of seasoning may be required for other types of firewood. The more humid the air, the better air circulation that is needed.

If you have problems getting the wood dry, remember that short logs dry more quickly than longer ones, because moisture exits fifteen to twenty times faster through the ends than through the sides. The second solution is to split your logs thinner to expose more surface.

Firewood from thin, unsplit trees or branches may give another surprise. Often they are still damp after one year's seasoning, whereas larger logs, chopped at the same time, burn easily. The explanation is that the exposed, breathing surface is small, relative to the length. The solution is to strip away a length of bark, allowing the moisture to escape more readily (see page 45).

Wood by Midsummer

An old Norwegian adage has it that wood should be felled, split, and stacked by Easter, and if that is done it will be dry enough for use by late June. By a curious coincidence midsummer is traditionally celebrated in Norway with a bonfire that is kept burning through much of the short night (the adage primarily refers to wood intended for heating in the coming winter). There is always the risk that such a calendar-based rule will strike modern ears as an irritating piece of obsolete country lore. But if we study the growth cycles of trees and compare them with the meteorological data, it soon becomes clear that there is good sense in the old saying.

Before Easter there is less moisture in the trees to dry out. By Easter the

A beautifully executed full-height stack—the master-class version of the Norwegian sun-wall woodpile, here covering all four walls of a house. Stacked by Inge Hådem, who won the Ark National Firewood Stacking Competition in 2012. This stack was made in the spring of 2014.

nighttime temperature often falls to below zero, meaning the wood is easier to split. But here is the clever bit: In the long and narrow land of Norway, the average moisture content of the air varies greatly from place to place. It is common in most areas, however, that around the time of the midsummer festival, the moisture content of the air rises dramatically, from about 55 percent to more than 80 percent. Morning dew begins to appear at around the same time, a sign that there is more moisture in the air. On average there is also less rain in the spring months. All these factors are significant, which is why spring drying is such a well-established concept among woodsmen.

If the wood is split into logs about fourteen inches long or less and stacked in good airing conditions early in the spring, the level of moisture can sink surprisingly fast from 45 percent to 35 percent within a couple of days, and down to about 30 percent by the end of the first week. After a full month in good drying conditions the figure can drop further, to about 22 percent, and after two months all the way down to 15 percent, a result that accords well with the adage. After that there is nothing to be gained from further airing and the logs can be moved straight to the wood cellar or shed. This is true for the Nordic countries, where the climate is moderately warm, and for wood such as birch and pine. In damp climates the drying process is slower; in hot climates, quicker. Most wood enthusiasts in the Northern Hemisphere countries make it a rule to get the process under way as early as possible in the spring.

Dry wood enters a state known as equilibrium humidity, meaning that its moisture content will reflect the moisture content of the air. In the late autumn, wood that has been stored outdoors actually gets damper as moisture is absorbed from the surrounding atmosphere. Yet the tree's equilibrium humidity will always be lower than the moisture content of the air, and it is slow to change. A moisture content of 60 percent at a temperature of sixty-eight degrees Fahrenheit (twenty degrees Celsius) gives an equilibrium humidity of around 12.5 percent, whereas damp weather over a lengthy period can raise the equilibrium humidity to around 20 percent. In Norway the air with the lowest naturally occurring moisture content is found indoors in Longyearbyen, in Svalbard, an almost polar archipelago of Norway, in the winter, where the normal figure is 23 percent. The moisture content of indoor wood paneling there will fall to 5 percent.

In the most common kinds of trees there is little difference in the equilibrium humidity. An exception is the alder, in which the moisture content can drop to as low as 8 percent under normal outdoor drying conditions. A good way of keeping track of the drying process is to mark four or five logs from the woodpile and weigh them at weekly intervals. When the weight finally stabilizes the logs have reached equilibrium humidity.

How Dry Is the Wood?

Everyone knows someone who has bought sacks of wood from a service station or garden center only to discover that the wood will not burn. So for professional wood merchants, it is a matter of pride that the wood they sell is dry, and they can prove it. They are only too happy to answer your questions about when the wood was chopped, and how it has been stored. Such information is actually the best way to make sure the wood you are considering buying is a quality

When wood from birch and many other medium-density deciduous trees is dry, it is possible to blow through it—made visible by smearing dishwasher liquid at the end. As well as a wood-cutter's party trick, it is also proof that wood dries fastest through the ends.

batch. Avoid wood with fungus or mold growing on it. It doesn't look good, but more important, these are signs that the wood has not been treated well. The best way to ensure good results is to buy the wood fresh in the spring and dry it yourself.

Good firewood should be hard, dry, and clean. The exposed surfaces at the ends of the log will begin to crack as it dries, and such visible signs are actually the most important indication that the wood really is dry. Another good sign is when the growth rings are slightly prominent. The scent is another useful indication. The aroma of sap and resin grows fainter as wood dries, and ultimately disappears completely. Exposure to a lot of sunlight will cause the wood to yellow. Old wood will turn gray if it gets wet and dries again. Very often the bark will come off.

A common method of discovering if wood is seasoned or not is to knock two logs together. If dry, there will be a hard, ringing sound, and if damp, a dull, unresonant thud. At best, however, this is only a rough guide. Electronic

moisture meters might seem a little fussy, but are a good alternative. They can be misleading, however, because the highest concentration of moisture is always to be found in the middle of the log, often close to the bark. Split a couple of logs down the center and take a reading there. Push the sensors well into the wood. Also note that the meters usually show humidity percentage as calculated for building materials.

However unlikely it sounds, another test that actually works involves smearing dishwashing liquid on one end of a log from a deciduous tree and blowing hard on the other. Dry wood is porous enough for air to pass through it, so soap bubbles should form at the other end. The trick is also a good proof that moisture exits most readily through the ends of the log, and a good standby when confronted by skeptical owners of fan heaters—every woodsman should have it in his repertoire of party tricks. Unfortunately, wood that is not completely dry will also produce small bubbles, so the method is not infallible. The cellular structure of conifers is closed and the trick will rarely work with these unless they have large drying cracks. With wood from moderately hard deciduous trees, it is possible to produce bubbles from a log more than sixteen inches long.

In the end nothing beats the simple dry-and-weigh method used by the professionals at the Norwegian Institute of Wood Technology (which has an advisory body known as the Drying Club). Results obtained here are at the scholarly level, and if you quote them you may well score heavily in the competition to find the biggest nerd on your block; your compensation is to know that you are unshakable in any discussion on the subject of what really does affect the amount of moisture in the wood.

The first step involves making a precise record of the weight of the logs. They are then split into sticks weighing about one hundred grams or less; otherwise, the experiment will take an eternity. Each "log" is then loosely bound again and its weight recorded before being placed in an ordinary kitchen oven at 212 degrees Fahrenheit (100 degrees Celsius)—the ideal temperature is 217.5 degrees Fahrenheit (103 degrees Celsius). Anything above this may affect the dry substance in the wood and produce inaccurate results. The oven fan is turned on to circulate the air, and the oven door left slightly open to let the moisture escape. The fine aroma of warm wood soon graces the kitchen.

At regular intervals the control "logs" are removed and weighed, and their weights recorded. A computer spreadsheet is ideal for the task. The weight loss

OPPOSITE The dry-and-weigh experiment can be carried out in an ordinary kitchen stove. It makes a lovely aroma in the kitchen but is messy and best done when your partner is away from home.

will be less and less, and when readings from successive weighings show no decrease at all, the wood has reached zero percent moisture. In practice there may be a tiny residue, but the zero reading is adequate for our purposes. The weight records the amount of dry substance in the wood, and a comparison with the original weight will show the percentage of moisture present in the logs at the start of the experiment. If the logs are weighed while still green and again during the course of the drying period, this will give a percentage for the amount of moisture in the tree at the time it was felled, and allow comparisons to be made with the different methods of drying and the influence of the weather on the process, with the ultimate aim of choosing a method that suits the type of wood at your disposal and the characteristics of your local climate.

Drying in the oven will take several hours, perhaps even a day or two, depending on how wet and chunky the wood is. The best results are obtained if samples are taken across the whole diameter of the tree, since the outermost wood has a higher moisture content than heartwood from the same tree. The logs should be an assortment typical of the whole woodpile.

How Much Heat Will I Get?

One kilogram (2.2 pounds) of completely dry wood from any type of tree contains a potential 5.32 kilowatt-hours of energy. This is the equivalent of 8600 Btu/pound. Whether one burns one kilogram of offcuts or prime-quality oak, the amount of heat produced will be the same. (The main exceptions are birch bark and birch twigs, which give 20 percent more.) In practice, the result is lower than 5.32 kilowatt-hours—the actual figure is closer to 3.2 kilowatt-hours for every kilogram. Two factors account for the discrepancy: the moisture content of the wood, and the fact that the stove is not able to utilize all the energy. A 20 percent moisture content in the wood is, by itself, enough to reduce the figure to 4.2 kilowatt-hours for every kilogram (in actual fact, the value goes down by slightly more than the percentage of moisture for technical reasons connected with the combustion process).

The other source of loss is the woodstove itself. No stove is capable of using all the energy in the wood because the gases do not combust completely and a certain amount of heat is lost to the chimney. The efficiency level in older stoves without clean-burning technology is somewhere between 40 percent and 60 percent, and in modern, clean-burning stoves between 60 percent and 80 percent. In an open fireplace it can be as low as 10–15 percent. The figure used in the trade for a typical modern Norwegian stove is 75 percent. That leaves us with a residue of approximately 3.2 kilowatt-hours for every kilogram of wood. That same wood burned in a stove with an efficiency level of 40 percent would yield only 1.7 kilowatt-hours, and in a large open fireplace with just a 15 percent efficiency, almost no heat at all would be generated—a mere 0.6 kilowatt-hours. With almost perfect conditions—some direct-fired boilers are close to 90 percent efficient—and wood with a moisture content as low as 13 percent (which is, realistically, the best that can be achieved with wood seasoned outdoors in a dry summer), the figure can be as high as 4.1 kilowatt-hours for every kilogram of wood.

Burning wood that is still almost green produces even worse results. More of the energy goes to driving out the moisture, and the heat generated will be so feeble that for most of the time the stove will not even reach normal working temperature, no matter how well designed it is.

Old Wood

A commonly held belief is "old wood, scant heat"—in other words, firewood that has been stored for a number of years loses heat energy. Research carried out at the Norwegian Institute of Wood Technology shows that twenty to thirty years' storage of dry wood does not affect its heating value to any significant extent. Certain volatile elements and oils will evaporate in the course of time, and pine is the most prolific producer of volatile elements, especially terpene and hexanal. (Terpene is what gives pine paneling and log cabins their characteristic smell.) However, firewood contains around 6 percent hydrogen, and this is a very important component, as it gives one-third of the total heat output. This hydrogen will gradually vanish, but sixty to one hundred years must pass before there is any notable difference in the heating quality.

In most cases the explanation for poor performance is that in the course of its long storage the wood has become a bit damp, or that the original phase of the drying process was slow (see page 46). If the moisture content has been in excess of 25 percent for some time then rot will cause the wood to degenerate; even if the percentage drops below this, the process will continue to reduce the heating value because of the internal fungus. Such wood is a huge disappointment in the stove—it looks normal but burns reluctantly and provides little warmth. Over time degeneration as a result of this kind of saprophytic fungus will cause the wood to lose weight. Other types of fungi do not break down the tree, but rather attack its outer layers and make the wood more porous to atmospheric moisture. These can remain active with moisture levels down to 20 percent.

EG GREV NED MIN ELD
SENT OM KVELD
NAAR DAGEN ER SLUT
GUD GJE MIN ELD
ALDER SLOKNA UT
JØTUL 1940

CHAPTER 7

THE STOVE

You can always see a face in the fire.

—Henry David Thoreau, *Walden*, 1854

People forget their schoolmates. They forget their holidays and their favorite toys, but they never forget the woodstove that warmed them in their childhood homes. The only thing that could thaw a frozen twelve-year-old who'd been out playing the whole of a February day. The same memory every evening: the blue darkness descending over winter snow as the thick woolen socks hung up to dry dripped and hissed onto the woodstove. Not too close, and not too far away. Sit there and absorb the heat from the glowing fire, with the air vent open full and the tongues of flame playing across the wooden floor.

There are a lot of woodstoves in northern Europe—in Norway alone, 1.2 million households have at least one. The cast-iron stove is the most widespread design used in Norway and Denmark, and that has been the case for centuries; Sweden and Finland prefer stoves that are heavy and have to be built as an integral part of the house, but are better at storing energy.

A wide range of stoves can be found in Scandinavian homes. Some are dour and utilitarian, others an elegant synthesis of the convenience of bioenergy and modern interior decor, still others are cast-iron treasures that tell a fascinating tale of industrial history. During the period when Denmark and Norway shared a king before 1814, Norwegian ironworks held a monopoly on the sale of iron and produced stoves adapted to the needs of every layer of society. In former times there were foundries making stoves in every county in Norway, for the simple reason that the obstacles to transporting something as heavy as that around the country were formidable (though honorable mention must be made of a case from Sweden—the pioneer in Mikael Niemi's novel *Popular Music from Vittula* carries not only a woodstove but a pregnant wife across two counties on his back on his way to settle in Norrland). For previous generations the need was for something that warmed well at a reasonable price; today's stoves are just as much a part of the interior decor of the house—and people are willing to pay for style. The advent of clean-burning stoves in the 1990s saw the start of a widespread revolution in heating culture in Norway and Denmark, with Sweden and Finland following a little later. The figures show that more than half the firewood used in Norway now burns in clean-burning stoves. Design has clearly played its part in this environmental advance, with the aesthetic beauties of the new range of stoves accelerating the speed of changeover from the old-style burners.

The marketing people in the main stove manufacturers and importers realized some time ago that the last word on what stove to buy always comes from the most aesthetically aware member of the household. According to a marketing survey carried out in 2002, the sequence of events in most sales was remarkably similar: The man of the house would be sent to find a stove. When he came home and presented his choice it would be vetoed by the house aesthetician. The wife would then drive out, examine the available choices herself, and pick another one. The survey showed that nine times out of ten the choice was made by a woman.

Clean-Burning Revolution

The difference between an old-fashioned and a clean-burning stove is that the clean-burning stoves use an extra supply of heated air. This ensures that the gases will ignite and combust even when the stove is fired only at a moderate rate. Roughly speaking, an old-fashioned model will make use of 40–60 percent of the potential energy in the wood, whereas a clean-burning model will use up

PAGE 146 The Jøtul 118 with the famous "magic spell" from medieval times. See page 168 for translation.

PREVIOUS LEFT Norway is a world leader in the field of stove design and combustion technology. This is the Kube 5, a soapstone stove by Granit Kleber of Otta, in Gudbrandsdalen. Clean-burning technology is allied to good heat storage.

PREVIOUS RIGHT Drafn was one of many fine kitchen stoves made at the Drammen Iron Foundry. This is a 33H from the 1920s, immaculately restored by Helge Løkamoen from Notodden, who runs a crisis center for old stoves.

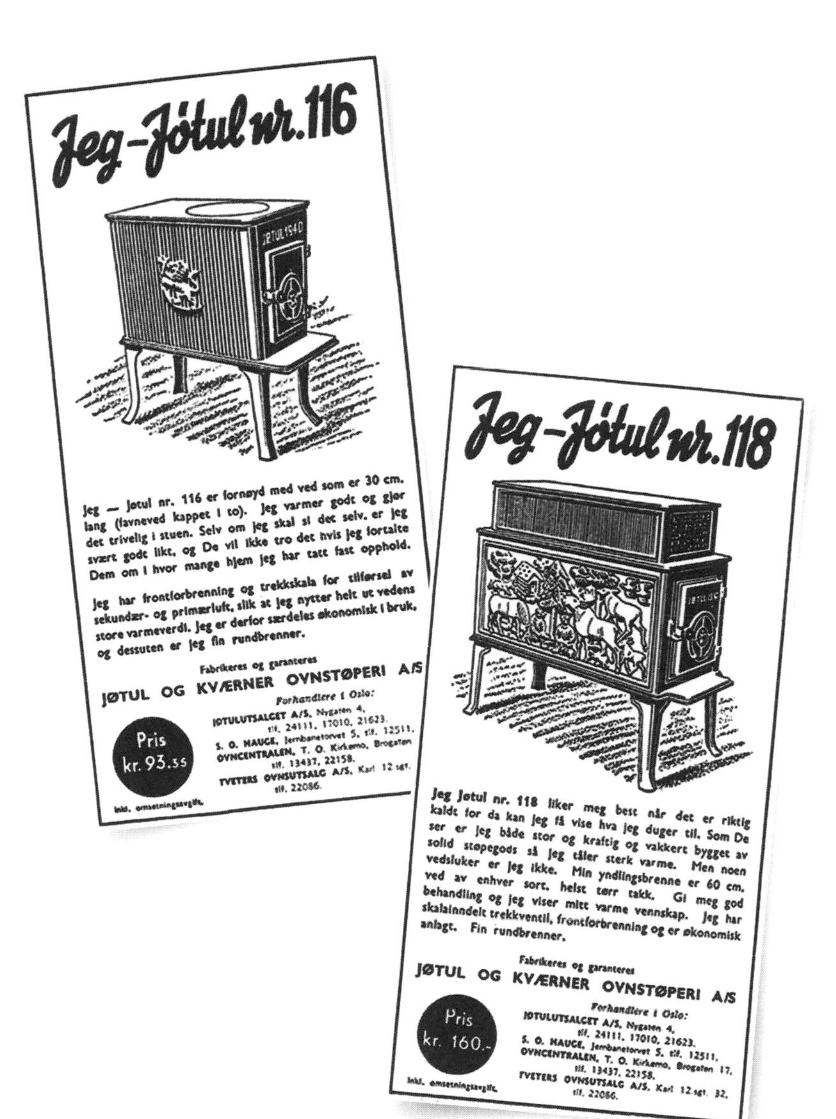

The Jøtul 116 appeared on the market at the outbreak of the Second World War and was later celebrated as no. 602, the world's most extensively manufactured stove. A peacetime variant, the Pax 1945, appeared shortly after the war. After a short break in production, classic stove no. 118 is again available, now with less decoration on the side panels. The advertisements are from *Aftenposten*, January 22, 1941.

Many Norwegians know the verse inscribed on the side of no. 118 by heart (see page 146). This is a "magic spell" from medieval Norwegian folklore. It was originally recited in the evening, when people raked the glowing cinders together and covered the fire for the night. The verse was also important on Eldbjørgsdagen, January 7, the traditional Christmas "fire day," on which a sacrificial feast was held with the aim of protecting the house from fire in the coming year. The tradition was observed in parts of Norway until as recently as the 1920s.

around 60–80 percent. In the old-fashioned stove a lot of the gases disappear up the chimney before combusting, whereas the modern stoves burn up almost all the gases and particles, converting them into heat and not pollution. Such ovens also use less wood. Some have an efficiency rate as high as 92 percent, and the new technology delivers a major reduction in the amount of atmospheric dust emitted. As of this writing (2015), some stoves have an emission rate as low as 1.25 grams of particle pollution for each kilogram of wood burned.

With the current focus on clean burning, it is worth bearing in mind that this has in fact always been a goal for the manufacturers, for the simple reason that the cleaner a stove burns the better the use it can make of the potential energy in the wood. And the more heat a stove generates, the better it will perform in the marketplace. The old-fashioned stoves are able to burn up most of the smoke gases too, but only by firing hard and with the draft fully open. Some of the older stoves (such as the Jøtul 602 and the 118) that have smoke combustion chambers can burn almost as efficiently as a modern stove under these conditions; unfortunately, they continue to give off some particle pollution, even though this is greatly reduced when firing intensely.

The Stove of the Future

SINTEF (the Foundation for Industrial and Scientific Research), with headquarters in Trondheim, in Norway, is a world leader in the field of combustion research and plays an active role in the development of Norwegian woodstoves. SINTEF's view is that advances in the field are directly related to social change. In former times it was normal to have large households made up of three generations: children, parents, and grandparents. It was the duty of the grandparents to see to the heating. In our day the household consists of parents and children, with the older generation living alone or in a home for the elderly. Houses tend to be unoccupied during the daytime because people are out at work, at school, or in the nursery. The incidental effect of this social change is that houses are cold in the middle of the day, unless the temperature is maintained by electricity, pellets, or heating fuel. Modern houses are also much more efficiently insulated against the cold. The old black woodstoves actually burned quite cleanly when fired hard, which was in any case necessary because old houses were cold and drafty. But hard firing doesn't suit a modern house—after a while it heats too much. SINTEF's research shows that a lot of the pollution from woodstoves comes about because people adjust the heating by closing off the ventilation, frequently in stoves that are slightly oversized. Pollution occurs, and a significant amount of energy in the wood remains unexploited.

The clean-burning stoves offered a solution to this problem, and they were often also a little smaller. Efforts are now concentrated on designing a new generation of stoves that take account of the change in the composition of a typical modern family, modern work patterns, and a low-energy house that is usually unoccupied during the daytime.

Prototypes already exist of stoves able to feed themselves automatically, so that the logs burn only at the end, like a cigarette. Other possibilities include hybrid stoves that can burn both wood and pellets and so keep the house warm without the need for constant attention. The odd fact is that for the first time in history, manufacturers are looking for ways to make a stove that gives off *less* warmth—one that burns efficiently on a low load, and where each separate feeding lasts a long time. There are, of course, a number of well-established ways of storing heat, either in solid form or in water, and interest in these technologies is reviving as a way of dealing with the problem of how to maintain background heating in an extremely well-insulated and energy-passive house.

Installation and Maintenance

A long flue pipe will increase the heating efficiency of a stove by about one kilowatt per meter. Another inexpensive improvement is the use of electric fans to circulate the warm air. Ceiling fans with large blades can move a lot of air without making too much noise. Smaller fans should be placed close to the stove. The Canadian company Ecofan makes models that work without electricity—they sit on top of the stove, and the heat powers a thermoelectric element that starts the fan motor turning.

Enthusiasts who want a reliable guide to the efficiency of their firing method invest in a stove thermometer. These measure either the surface temperature or the temperature of the flue gases. Note that certain types of thermometers are designed to measure the temperature of the gases inside the stove, and the scale will therefore show roughly twice the temperature of the surface on which the thermometer is mounted. The surface temperature can also be measured with an infrared thermometer. This can provide surprisingly good insight into the functioning of the stove and its optimal wood load.

In all stoves there is an ideal draft that provides maximum efficiency, and the most common complaint about stoves is that they do not fit the flue. An exhaust fan may sometimes help if the updraft is too weak, though too strong an updraft will also cause a problem, in the form of reduced heat. Long yellow flames licking up toward the flue are one sign that this is happening. The forces of nature are malign in this instance, because the draft increases in proportion to the decrease in the outside temperature, and it is on just such cold days that the heat vanishes up the flue. One solution is to install a damper in the smoke

pipe that can be adjusted manually to reduce the draft by between zero and 70 percent. Leave it wide open when lighting the fire, and then open just enough to keep the flames rolling along nicely inside the firebox, ideally with a bluish tinge at their center. The heat gain can be very considerable.

The most important maintenance you can carry out on the stove and the inside of the flue is to clean them regularly. Soot is such an efficient insulator that a layer just 0.01 inches thick imposes a 10 percent heat loss because the heat is not transmitted out through the iron but drawn away through the flue. Half an hour with a wire brush and an ash vacuum cleaner can bring a dramatic improvement in efficiency; in Trondheim the Institute of Chimney Sweeps found that, on average, the improvement was 30 percent. In former times many Norwegian army barracks were heated using wood, especially in Finnmark, and a soldier's duties included regular cleaning of the insides of the stoves. Although a lot of soldiers thought it a meaningless and petty imposition of disciplines the purpose was, in fact, to ensure that the stoves continued to produce good heat.

At the same time the stove should be examined for cracks. Loose or damaged stove rope should be changed or repaired with glue to ensure that the only air entering the stove is doing so through the proper avenues. Stoves without an ash pan should have a layer of crude sand at the bottom to prevent the base plate from cracking.

A good chimney sweep will give you useful and objective advice on the condition of your stove and the best way to use it. He will be able to measure the draft and the condition of the chimney, perhaps by the use of an inspection camera.

Types of Stoves

The Open Fireplace/Hearth The original domestic heat source. It let man bring fire indoors, and to this day nothing brings one closer to the magic of the flames and the radiant heat than the hearth. In masonry buildings, the fireplace and chimney may be an integral part of the structure, usually forming a short wall. In Scandinavia, however, wooden houses are more widespread. For this reason the fireplaces are usually made from stone and placed in a corner, connected to a separate chimney of concrete or brick. When conifers are burned, a spark arrestor made of fine mesh is used, its pattern often forming shadows and silhouettes against the flames.

Unfortunately, the open fireplace has significant drawbacks. Most of the heat

OPPOSITE The genius of bygone times: The tiered stove offered a large surface area that radiated a generous amount of heat. This one is a Kragerø no. 45 from about 1900.

is lost through the chimney (in Norway such energy losses are called "heating for the crow"). It is also far more difficult to achieve satisfactory combustion when the air supply cannot be controlled. Thus open fireplaces will release more smoke and soot, and in some areas they are prohibited. It is possible, though, especially with the method of stoking from the top (see page 172), to reduce emissions greatly. One should tend the fire often, and ensure that it has distinct flames or embers and is not left smoking. It is a good practice to leave a thick layer of ashes in the fireplace, since it will conserve the heat of the embers for a long time.

The Closed Iron Stove The most common type of woodstove in the world is the rectangular, box-shaped cast-iron stove with a door on the short side. Even small stoves of this type give off a lot of heat and are usually quite easy to light. Since they are flat on the top it is also possible to cook on them, and some are equipped with a cooking plate of turned iron. With such a stove in the house you will comfortably be able to survive a power outage, even one that lasts a few days. It is often in a crisis that the beauties of the simple and pragmatic appear most clearly.

The best-known Scandinavian makes are the Danish Morsø 2B and the Norwegian-made Jøtul 602. Both are still in production, and these days in a clean-burning design. More than a million 602s have been produced, which is a world record for woodstoves. This small, square classic with its round air intake is one of Norway's most successful exports. The design was copied in both Japan and Taiwan, and Jøtul spent years pursuing the pirates through international courts of law. The motif on the side is the Norwegian heraldic lion as it was in 1844, during the reign of Oscar I, king of Sweden and Norway. The lion holds an ax with a handle so long and curved it is able to stand on it itself.

Iron Stoves with Glass Doors One of the drawbacks of the classic black-box design is that you are denied the aesthetic pleasure of watching the flames. Another drawback is that it is not easy to tell when the fire needs refueling, whether it is dying out or is smoldering and polluting. For these reasons, by far the most popular stoves on the market today are tall models with curved lines and large glass doors. Another major advantage of these is that infrared radiation through the glass doors is transformed into heat as it encounters the body. All the major manufacturers in Norway and Denmark concentrate their development money on improving this type of stove, competing with each other to achieve the most successful combination of good design, low emissions, and maximum efficiency. Good design is actually encouraged by the environmental lobby, who recognize that the more aesthetically attractive the design is, the more likely people are to switch to clean-burning stoves. The shape of the stoves means that the maximum length of log they can take is rarely more than twelve inches, but their efficiency more than makes up for this inconvenience.

Soapstone Stoves Soapstone is found almost everywhere in Norway, and has been used in making fireplaces for more than three hundred years. Stone in general is good at storing heat, and soapstone is outstandingly good. Its porosity means that it retains heat four times longer than granite. Large soapstone stoves can keep a house warm for several hours after the fire has gone out, their ability to retain heat depending on the thickness of the stone. With larger stoves the floor may have to be reinforced. The Granit Kleber company in the Gudbrandsdal valley (Oppland County) manufactures a range of elegant models designed to give particular prominence to the lovely surface of the soapstone.

The Kitchen Stove Until well into the 1950s a low kitchen stove with a built-in oven and heating plates that also kept the kitchen warm was standard in Scandinavian homes, and those stoves that survived the era of the electric radiator and the electric fire have decades of hard use etched across their iron casings. The shape is usually broad and deep, with room for large cooking pots, and the stoves have an ash pan that can be removed and emptied while the stove is in use. The hotplates usually have loose rings to improve the effect. The thick casting enables the stove to maintain a steady heat. The fuel used is usually finely chopped aspen or spruce, giving good control of the temperature.

These stoves were not toys, but trusty, functional installations that were the beating heart of a household. Even today blasé possessors of induction stoves might be surprised at their efficiency: large pots of water boil at remarkable speed, and stable temperatures make cooking and baking easy. Kitchen stoves are made by, among others, the Swedish firm of Josef Davidssons Eftr. In Italy, La Nordica makes a number of large ovens in a contemporary finish.

The Tiered Stove The tiered stove was once enormously popular for the ingenious way it solved the old problem of how to extract all the heat from the stove before it disappeared up the chimney. The solution was to guide it through a labyrinth that gave the smoke a larger contact surface to heat up on its way out. Ever since the first tiered stoves appeared on the market in the early eighteenth century, these have been an elegant presence in the home. Form and function complement each other superbly, and from the very beginning the foundries did their utmost to exploit the aesthetic potential of the design. The stoves were frequently 6.5 feet in height, and every visible surface would be decorated in some way. The channels for the smoke consisted of a number of relatively thin plates, often with open bars between them—a dream for the creatively inclined foundry. All tiered stoves are considered cultural treasures, and it is therefore still legal to reinstall an old one even though it is not clean burning. An export license has to be obtained if these are to be sold abroad. Today the only model made in Norway is the Ulefos no. 179, which has either two or three tiers.

The Tiled Stove The aura of culture and refinement surrounding these stoves, which are a Swedish institution, is not wholly a matter of aesthetics and comfort—the principle behind them is itself an almost 250-year-old stroke of genius. The way in which they work has acquired a new relevance to the problems of a modern, well-insulated house that stands empty for most of the hours of daylight: The heat is stored and released gradually, continuing also after the fire has gone out. The principle is similar to that of the tiered stove, but these stoves are made of brick and have a considerably longer exit route for the smoke. The outer surface is covered in ceramic tiles and affords limitless scope for decoration. The stove was developed in Sweden in the eighteenth century, with the country on the verge of a wood crisis. Besides revolutionizing the way wood was used, it gave people more time for other things, and in so doing meant that the hearth ceased to be the only feasible focal point of the household in winter. The interior smoke channels are twelve feet or longer and made of brick, a material that retains heat well and releases it slowly once the fire has gone out. After a break of almost eighty years the Swedes are once again actively working on improving the basic principle, recognizing that these stoves are ideal for the modern, well-insulated home. The design means that they do not need refueling more than two or three times daily. The heating curve of the Swedish Cronspisen reaches close to three thousand watts after six hours, and the makers have managed to get a thousand watts out of it a full twenty-four hours after the flames have died out.

The Finnish Mass Oven The pride of Finland, this is a variation on the tiled stove. It is heavy, often weighing more than three and a half tons, and often has to be built on-site, at the same time as the house. It is made of brick and clay, and the principle is the same one that is at work in tiered and tiled stoves, with the bricks being heated by the convoluted passage of the hot air through several levels as it makes its way up to the chimney. The typical Finnish mass oven, however, has *two* chambers, of which the inner one burns only the smoke gases.

OPPOSITE Modern stoves with clean-burn technology, like this Jøtul 473, will turn otherwise polluting gases and particles into heat. As seen in the photo, the firewood is burned with the top-down method described on page 172.

JØTUL

The stoves are extremely efficient and the best of them will be almost as warm to the touch a full twenty-four hours after the fire goes out. The disadvantage is that they are expensive to build and cannot be moved; but they use little wood for the amount of heat generated, and a single daily refueling is often enough to keep the whole house warm. These stoves are often built with oven compartments for baking bread and pizza, either as separate chambers or as part of the inner combustion chamber.

The Masonry Oven This is actually a stove designed especially for cooking and baking, but it is included here because it closely resembles the Finnish mass oven. Ordinary masonry bricks are used, but the principle here is that the cooking is done in the same chamber that the wood burns in. Hard firing heats the stone until it is burning hot and free from soot (the walls need to be heated until they are a gray-white before the stove can be used), then the ashes are removed and the bread or pizza placed inside. In large baking stoves the stone retains the heat long enough for several trays to be baked. Another good excuse for including this type of stove is that it is possible—for the fun of it, or in an emergency—to bake bread or pizza in an ordinary modern iron stove. The ash needs to be emptied out first, and it is a good precaution to give the inner walls a brush. You then add wood and start the fire, burning from the top down (see the next chapter) and preferably using wood from deciduous trees, since this burns more intensely. The heat should be so fierce that it burns away all the soot and ash in the firebox. As soon as the flames have died out completely you rake the glowing embers more or less flat and place firestones on top of them. These are normally about four by eight inches and one inch or more thick—just check to make sure you have the nontoxic kind. By the time the stones are really hot the embers will have collapsed. Use oven gloves to adjust the stones so that they are lying more or less level, and then put whatever you want to bake on top of them. The embers need not have gone out completely because by this time they will be past the stage of giving off smoke. Pizza and rolls can be cooked surprisingly quickly and well using this method.

Boilers We see a quantum leap in the exploitation of the energy potential of wood when it is used to heat liquids. Unfortunately, air is a rather unreliable conductor of heat, being difficult to force in any direction other than upward. Water is more amenable and the advent of central heating toward the end of the nineteenth century began a revolution in ways of heating multistory buildings. The technology even inspired new directions in architecture. The principle was the same as that used today: A powerful source heats up a large reservoir of liquid, which is then pumped around the house to the radiators or underfloor heaters.

Many boilers run on wood, and these optimize the combustion process in a completely different way from the standard woodstove. All modern systems have gasification—the wood is heated to a controlled temperature and the gas burned in a separate chamber. (This actually utilizes the same principle as the wood-gas vehicles used during the Second World War, with a large boiler mounted on the back of the vehicles filled with small pieces of aspen or alder. The chips were kept smoldering and the gases led through a pipe to the intake manifold.) Gasification was for many years a sleeping technology—old boilers used single-chamber combustion, like an ordinary stove—but it is now being used in combination with modern gas and temperature sensors that monitor the temperature and the smoke gases. The wood is heated to a controlled temperature and the gases forced through nozzles to a secondary chamber where combustion takes place. Air vents and extractor fans are adjusted automatically to give the best and cleanest possible results. Boilers intended for private use are typically designed to need one or two refuelings every twenty-four hours in the winter, but larger water tanks can store the heat for days at a time. Some models send status reports to a control panel in the living room that indicates when it is time to refuel. Many boilers have storage bunkers with room for two hundred liters of wood or more, some have automatic feeding, and the usual length of log is between twenty and forty-five inches. Some enthusiasts make it a point of honor to use less electricity in the winter because their boilers eliminate the need for electrically heating the hot-water tank. A number of different types of boilers are made for use in very large buildings, and such installations can mean a more prominent place for wood in a modern, green economy.

OVERLEAF Arne Odmund Herstadhagen doesn't get cold in the winter. He starts working on his wood early in the spring. Short or non-standard-size wood is stored in the drying bins. The remaining well-shaped and standard sizes make good building blocks for the solid and handsome cord stacks in front.

CHAPTER 8

THE FIRE

I open the wrought-iron door, where the birch logs lie curled like red and gold snakes, the small green and blue flames hissing up from the ashes. It strikes me that this way of relaxing, for the old as well as the young, is such an ancient way, perhaps the most ancient of them all. We've been sitting like this for hundreds, thousands of years.

—Ingvar Ambjørnsen, *Opp Oridongo*, 2009

Here it comes at last. The cold time. The great time. It might be a day in late October. You've had the fan heater on full and still the house feels chilly. You've been out driving with the heating control turned up high the whole time. That's when it hits you: Winter's here. Time to take a stroll out to the woodpile and get started.

A moment of truth arrives just after Christmas: Have you got enough? For a man may skimp on the price of his daughter's confirmation and still be forgiven; he may forget, and forget again, to order new furniture for the garden; he may even prefer to build a new garage rather than take his family on a Mediterranean holiday. But for the man who would see his family freeze in winter, there can be no forgiveness.

As autumn finally gives way to winter and the first match is about to light the first fire of a new season, a moment's reflection may be in order. Each time we hold a match, what we have in our fingers is one of mankind's most crucial inventions. For thousands of years the power to keep flame alive was of critical importance. Small wonder that fire has a central place in almost all religions. It was fire that protected from the cold and the dangers of the night. Those who stayed close to it were rewarded with warmth and security. Light was good, darkness bad. And yet fire, like some hard god, was the possessor of a vast and unquenchable power.

All people in the world have made use of fire, but up until the nineteenth century there were still a few tribes unable to *start* a fire. They got it either from other tribes or from forest fires caused by lightning, and they were careful to preserve the embers. Fire was precious, and many cultures have a tradition of keeping it alive. The longest continually burning fire is probably that in the Zoroastrian fire temple in Yazd, in Iran, whose flame has been tended ceaselessly since 470 CE. The fire is fed by priests, using the wood of the apricot and almond trees. It has been moved three times: in 1174, again in 1474, and to its current location in 1940.

Firm rules governed the relationship with fire in old Norwegian folk beliefs. A Norwegian community statute book from 1687 warned that, should anyone pass between a pregnant woman and the fire by which she was warming herself, the child would be born with a squint. One of the most dramatic celebrations of fire is actually Norwegian in origin, the festival of Up Helly Aa, on the Shetland Islands of Scotland. This is a heady mixture of a post-Christmas party and a commemoration of the islands' history as a former part of the kingdom of Norway. At its climax small Viking ships are launched onto the sea and set ablaze. So extreme are the aftereffects of the business, with its battles, its Viking helmets, parades, singing, and drinking, that the day after is an official public holiday throughout the islands. Shetlanders also preserve a tradition of taking to a new home the glowing ashes from the stove in their old one.

An inscription on the classic Jøtul 118 stove, which is exported all over the world, is another reminder of how important wood has historically been to Norway. The inscription reads: *Eg grev ned min eld sent om kveld. Naar dagen er slut, Gud gje min eld alder slokna ut* (I damp down my fire, late at night, when day is done. God grant that my fire never go out). This fire prayer is from the Norwegian Middle Ages and was originally said last thing at night, when the fire

PREVIOUS LEFT There's bound to be a good supply of matches in this mountain cabin, in Gullhaugen in Fåvang.

PREVIOUS RIGHT A campfire that burns well will not give off much annoying smoke. A good procedure is to ensure that the hottest spot is on *top* of the campfire.

In former times, kitchen stoves for wood were found in every Norwegian home. Today, the Italian La Nordica stove provides the same pleasure and efficiency. It will heat even a large room or a cabin, as well as provide hot water and constant heat for all sorts of cooking.

was damped down. A folk belief of the time held that fire kept the forces of evil at bay, and should the fire go out, these might yet enter the house. The verse also played an important role on Eldbjørgsdagen, January 7, "fire day" in the old Norwegian calendar. This was a sacrificial feast intended to stave off the danger of house fires in the new year. Both the fire prayer and the celebration of the day continued to be practiced in Norway as late as the 1920s.

The Combustion Process

Lighting a fire might seem a simple enough task, but it can be the source of surprising difficulties, swathed in smoke and frustration. Failure can give rise to a disheartening sense of defeat, which can in some cases assume irrational dimensions—possibly an echo of primitive man's anxiety at the thought of being without fire. Acquiring an understanding of how the combustion process actually works should make it possible to succeed every time, always provided that the wood is dry.

Basically, three phases are involved. When the log is put on the fire the first thing that happens is that the moisture inside it begins to evaporate. This takes energy from the rest of the fire, and is the reason unseasoned wood actually steals your heat. Once the outer parts of the log are dry enough the temperature

Woodsmoke gases are not exhaust gases. They are rich in energy, and when combustion processes are fully engaged all that goes up the chimney is a small amount of steam and pale smoke. The rest is turned into heat.

will rise, the wood will release gas, and flames will appear at the second stage of the process. It may look as though the wood itself has caught fire, but in fact something rather surprising is happening: At this stage the flames we see are not from the wood itself, but from burning gases that jet from within it once a certain temperature has been reached. To put it simply, the wood is cooked until it is dry. Gases start to percolate out and ignite. If they are to burn vigorously they need a good supply of air. Once the gases in the wood are exhausted the third and final stage begins. The wood has by this time become charcoal and glows at a high temperature. Much less air is needed by now. In practice all three processes go on at the same time, simultaneously fostering each other.

Lighting the Fire

A match alone will not make a big log catch fire. All fires have to grow, and the process should always be carried out in three stages, with three materials.

Purists swear by the same procedure for lighting a woodstove as for starting a bonfire: birch bark first, then dry twigs, and when enough heat has developed, add the larger logs.

A classic, simple Scandinavian method is to use crumpled newspaper for step one—any will do, though romantics prefer a sun-browned local edition with a half-finished crossword puzzle still showing. Magazines are of no use for the job because glossy paper contains 50 percent industrial minerals—in a word, stone. It burns very poorly and produces a large quantity of ash.

The drawback of using newspaper is that you have to arrange the kindling *before* you start the fire, and burning paper also requires a lot of oxygen. Small firestarter briquettes or solid fuel tablets can burn for several minutes, and make for a better, controlled procedure because you can lay kindling on top even when the briquette is burning. They make little heat, but their long burning time makes up for that.

Spruce or a similar light wood is perfect for kindling, because it is easy to split into thin sticks, even with a knife. Once the kindling has caught, the heat will be enough to get bigger logs going.

The Valley and Bridge

A versatile and effective way to start a fire is the so-called valley-and-bridge method. Two logs are laid side by side to create the valley. Then newspaper or a firestarter is squashed in between. Thin sticks of kindling are laid across to form the bridge. This works well because the burning bridge will stand above the ashes and air can circulate around the flames.

Variations exist. A good one is to put a thin log on top of the bridge; when the bridge has burned out and collapses, this burning log will fall into the valley. This procedure is an excellent combination with the stoking-from-the-top method described below.

Turbulence is important when lighting a fire. When the temperature is low, oxygen is reluctant to engage with the molecules in the fuel. Air in turbulence means that the fumes are bombarded with oxygen and combust more easily. That is why the fire burns more intensely in a stove where the door is left slightly ajar. Some houses can be so well insulated that it is advisable to leave a window open when lighting the fire.

A simple and very effective tool is a blowpipe or blow poker, a metal tube some two feet long that you can easily make yourself. Crimp one end slightly; that will increase the speed of the air. This gets a fire going surprisingly quickly and is much more efficient than a bellows. Once you have experienced its effects yourself, it will become a much-valued tool.

A chimney needs a good updraft, and this is created because warm air is lighter than cold air. It means the air in the flue must be warmer than the surrounding air, so don't be shy about piling on the kindling when starting a fire in a cold stove or fireplace. On days when there are wide fluctuations in the temperature you might experience the phenomenon known as downdraft, when the temperature of the air in the pipe falls below that of its surroundings. This is what causes smoke to come billowing out into the room, and anxious householders call in the chimney sweep, afraid an owl has gotten stuck in the chimney. When this happens one thing worth trying is burning newspaper up inside the stove. Once the firebox and chimney are warm, the draft makes it easy to keep the fire going.

Stoking from the Top

Most people start a small fire, and then put firewood on *top* of it when it has started to burn well. But the gases will always rise, and if there are no flames at the top to ignite them, they will pollute instead of giving heat. There is a solution, namely to make the hottest spot on *top* of the burning wood. In 2010 a nationwide campaign was launched in Norway to get more people to fire from the top down, because it reduces emissions and has many other benefits.

The method goes like this: A layer of logs is placed flat in the bottom of a cold stove. Kindling is placed on *top* of these—the valley-and-bridge method is perfect. Soon the fire begins to eat its way down. Gas is given off as the logs heat up, but there will always be flames higher up to ignite them. The method works especially well in modern clean-burning stoves with a large door, and in open fireplaces such as campfires. It takes a little practice, but is very useful to master.

The top-down method also makes it possible to start a fire and keep it burning in one step. This is useful for certain modern glass-door stoves that tend to let smoke escape into the room when the door is opened to insert the bigger logs. With the top-down method, a person can simply start the fire, wait until the wood has burned down and the embers (which do not emit smoke) are glowing, and then open the door and add the next supply of wood.

How to Burn with Minimum Pollution

Fires smoke and pollute if they do not burn intensely. Gases will already be emitted from the wood at temperatures between 200 and 300 degrees Fahrenheit (100 and 150 degrees Celsius), but nearly 650 degrees (350 degrees Celsius) are required to ignite them. In this temperature gap the gases (the smoke) will get up the chimney unburned. Note how smoke disappears immediately as soon as a flame appears nearby. For this reason a hot stove will burn cleanest and best—all the pollution is simply turned into heat.

You will get a good indication of the state of affairs simply by going outside to check the color and smell of the smoke. When things are as they ought to be, there should be only a thin, pale emission from the chimney, and there should be no smell—all the potentially harmful gases should have been consumed and transformed into heat.

Firing from the top down is a good technique to use in modern woodburning stoves where the air supply is at the top. A dense magazine of firewood is ignited by a small fire built on top of it.

This holds true for all stoves and fireplaces, and with the best clean-burning stoves *no* visible smoke at all will emerge from the chimney.

A good general rule to reduce pollution is to start the fire from the top down, then keep the fire burning briskly. The stove should be fed with smaller amounts of firewood at regular intervals, with the air-control lever close to fully open so the fire gets a good supply of air.

Once the house is warm, the temperature in the room is controlled by the amount of wood being burned, not by closing the air vent. You can also use a softer type of wood with a lower heating value.

A phenomenon often observed is that a fire with a single log usually goes out easily. Two or more logs should always be added each time, as this will create more air turbulence and keep the process moving along.

Also, read the manual. A lot of people treat modern stoves as though they are the same as the old ones they remember from their childhood. As one disillusioned salesman put it: "The instruction booklet is probably the first thing people burn." A clean-burning stove will rarely get *really* hot if the door is always left slightly ajar. And as a general rule, if the stove has a start-up air-control lever, it should be closed once the fire is well established. In due course small flames should be seen burning through the holes of the secondary air supply, a sign that the cycle is functioning.

Firing too fiercely, so that the oven starts to give off unusual sounds, is dangerous and can damage both the chimney and the stove. The round-the-clock method in which wood is kept at a constant low smolder isn't a good idea either, and is also a huge source of air pollution. The economy of the method is a false one, because the valuable gases do not ignite, leaving instead a thick, dark deposit of soot in the flue and chimney. Although the stove may preserve a lukewarm heat over several hours, the total heat gain is lower than if the wood combusted in the normal way. For this reason the last wood of the night should always be sizable logs of hardwood that combust normally; even if the fire goes out, the house's own insulation will retain much of the heat. A good idea is to leave a fairly thick layer of ash in the stove. The embers will last longer, and in the morning the ashes will probably still be lukewarm. If you ready some kindling and newspaper the evening before, it should prove a simple matter to get the fire going again the next day.

Can Wood Ever Be Too Dry?

Back on page 135 we examined the old belief that wood can sometimes be "too dry." Experiments carried out by SINTEF have shown that, from a technical point of view, wood can never be too dry, and that the presence of any moisture at all reduces the heat potential of the log. However, SINTEF has found good documentary proof of a phenomenon that probably lies behind this story, something that happens when using dry spruce or some other lighter wood, particularly if the logs have been finely split. When a large load of wood is fed into a hot stove with a glowing bed of ashes and a low air supply, the slight delay of the initial drying phase is missed and the process moves straight to phase two, immediately starting to release large quantities of gas. The release can happen so swiftly that older types of stoves are unable to provide enough oxygen to burn the gases, so they disappear up the flue before they have a chance to ignite. Since gases account for half the heat content of wood or more, the only energy being used is that from the carbon burning. The rest goes—quite literally—up in smoke, and the heating value of the wood is greatly reduced. Seen in this light, there is some justification for the old belief, because where the wood is less dry the gases are released more slowly and more of them are ignited. But the blame is being incorrectly apportioned, because here the culprit is not the wood itself but the supply of oxygen and the truncated combustion time. In clean-burning stoves the phenomenon is rarely observed, because these are built to maximize combustion of the flue gases. The problem also arises where the firing is too fierce. Research carried out by the Jøtul company has demonstrated that the efficiency of a stove will actually decline if it is fired beyond the level for which it was designed, precisely because the gases do not have time to ignite. Jøtul's instruction booklet states the facts plainly: "Firing too intensely sends much of the heating energy straight up the chimney as hot smoke."

The Chimney Sweep: The Woodburner's Best Friend

Stay friends with the chimney sweep. The sweeping removes the lining of soot from the flue and in doing so reduces the risk of a chimney fire. The amount and type of soot are also good indicators of the quality of your wood and the efficiency of your techniques as a woodburner. There will always be a certain amount of dry soot, especially when you use pine, but shiny, glasslike splinters are signs of creosote and indicate that combustion is poor, and possibly also that your wood has been imperfectly seasoned. Such hard deposits greatly increase the risk of a chimney fire. The danger of this is greatest at the start of a period of cold, particularly in the case of those inexperienced users who suddenly start to fire very intensely following a period of several days in which only a limited supply of air has been entering the stove. Some years ago the Institute of Chimney Sweeps published a surprising set of statistics showing that in Norway, many chimney fires occurred on a particular weekend in the spring, around the time when tax returns were due to be filed. The explanation was that people going through their papers had a tendency to burn any unwanted documents in the stove, and the sudden rise in heat in the stove set fire to the soot in the chimney. That was then. In the Norway of today modern techniques and clean-burning stoves produce a minimal amount of soot, and tax papers are filed using the Internet, so the problem has disappeared.

The signs indoors of a chimney fire are unnatural rumbling and roaring sounds from the stove and the chimney. Outside, sparks or flames will be seen coming from the chimney pot, so that neighbors or passersby are usually the ones to bring the news to the affected household. A small chimney fire burning dry soot is usually a relatively low-key affair and can even have the effect of cleaning the chimney, but a fire in creosote is a different matter altogether. The temperature will rise to 2,200 degrees Fahrenheit (1,200 degrees Celsius) or more, and large flakes that tumble down can block the chimney stack or fall into the stove. The temperature can rise so high that the chimney cracks and the ceiling and ceiling joists catch fire. The best way to avoid chimney fires is to fire properly with dry wood.

Because chimneys can be made of a variety of materials, it is a good idea to ask the local sweep for advice on how best to proceed in the event of a fire. In the Nordic countries the general advice is to close all the vents supplying all stoves, including those not in use, to retard the supply of air to the fire. Then call the fire company immediately.

One last golden rule: When emptying ashes, bear in mind the old saying about lost love: "Ashes still glowing soon turn into raging fires." Always use a metal bucket.

Liv Kristin and Peder Brenden, from Brumunddal, build classic cord stacks with 24-inch logs. Peder handles the chain saw and Liv Kristin does the splitting and stacking, usually in the mornings, after mucking out the cattle.

BRUMUNDDAL: THE CHRISTMAS WOOD HARVESTERS

Every Christmas, Liv Kristin and Peder Brenden host a family gathering at their farm in Brumunddal, in southern Norway. But the attractions are more than just the brawn and the marzipan ring cakes. All the members of the family enjoy outdoor work, and in the week between Christmas and New Year's they head out into the forest with tractor and chain saw.

Come spring the wood is stacked full-length in cords—a classic Norwegian method that involves the wood being cut twice. This has a number of advantages, among them that the wood can be stacked in stable, freestanding rows several meters long.

But during Christmas week they enjoy just puttering about, and content themselves with felling the trees, and trimming and gathering them. With so many willing hands and a tractor at their disposal, it doesn't take many sessions in the woods before the family has felled enough trees to keep them going for the year. A few bracing hours in the birch wood does wonders for the appetite by the time family members are ready to sit down to their spareribs and the rest of their Christmas dinner.

"Harvesting in Christmas week is a tradition on the farm," says Peder. "It could easily be done some other time, but there's something special about everyone getting together and working together in a place that all the family feels an attachment to from childhood."

According to Peder and Liv Kristin, harvesting the wood in winter has advantages for the work rhythm on the farm for the rest of the year. The trees have less moisture in them, and the snow and ground frost make it easier to drag the big logs about. Harvesting in milder seasons would mean tractor wheels plowing up the ground and damaging the undergrowth. Mud would cake the bark, which in turn would blunt the chain saws and other tools. The end product would be an unsightly mess.

The winter harvest avoids all these problems. In January the timbers are dragged by tractor to a site close to the farm and there chopped into logs some two feet long, the size traditionally known as cordwood. Not until it has dried is it cut into one-foot lengths, ready for the stove.

Liv Kristin is in charge of the splitting and stacking, and the long woodpiles are an eloquent testimony to the efficiency and precision of her work. She splits only when temperatures drop below freezing and the timber is frozen and comes apart easily and neatly. The couple are in principle opposed to the use of a wood processor and use only a chain saw and a hydraulic splitter.

Preparations for each stack start with the laying of long rails on the ground to ensure a stable base and avoid contact with damp. Then comes the stacking itself—and here the advantages of longer logs are seen to the full. At each end of the stack a block of crisscrossed logs is built about 3.5 feet high. This starting block provides a firm and solid support against which logs can be piled. Using long logs makes it simple to stack steadily; for obvious reasons the work also takes less time than it would with shorter logs. Liv Kristin appreciates the fact that the stability of the cordwood means she can stack the logs with plenty of space between them, facilitating the circulation of air and the drying process.

Most of the wood is birch, but the couple make sure there is a small amount of aspen in the stack as well. They particularly like the calm way it burns, and the fine, strong light it gives in the fireplace. It is also a link with the farm's own past, when it supplied the matchstick factory at Nittedal with aspen.

All heating in the main house is done by a single woodburning stove. The house was built in such a way that heat from this stove—a Swedish Contura—is distributed evenly throughout all the rooms. A fresh-air intake has been added to the stove so that it doesn't use already heated air from inside the house. The only additional source of heat is an electric radiator in the bathroom.

By Easter all the wood has been split and stacked and the couple start on the spring work, leaving the long, elegant woodpiles to dry in the heat of the sun and the fresh air. No tarpaulin is necessary; the small amount of rain that falls at this time of year quickly runs through the spacious stacks, and the cordwood is sturdy enough to withstand a strong wind without collapsing.

"On a farm it's important to guard against unpleasant surprises," says Liv Kristin. "A collapsed stack in the middle of the spring work—or any other kind of strictly seasonal work—can be a real nuisance."

Drying conditions are so good that the wood is ready for the next stage in July, after the silage making. The wood is loaded onto a trailer and driven up to the farm, cut in two with an electric saw, and stacked in the woodshed. Since the wood is already dry enough for use it can be packed tightly to maximize the space.

With cordwood the amount of handling before the wood finally makes it to the shed is halved. That is one reason why, in former times in Norway, logs were delivered in a standard length of some two feet, with the final cutting left to the user.

Cutting the wood twice this way does not increase the amount of handling involved. In fact, the number of operations involved remains the same. Handling, splitting, stacking, and transportation all take much less time, simply because, in the initial stages, logs that are twice the length require half the number of operations.

Another simplification is that the diameter of the logs in the final stage is so small that the work can be done with a small log saw. This is an advantage if a lot of the wood is chunky and rough. It does, however, require a splitter that can take logs of considerable length.

"Sometimes people passing by will stop, notice we're using the old-fashioned measurement, and say: 'That's nice, I remember that's the way we used to do it in my day,'" comments Peder. "But we don't do it for nostalgic reasons. It's just that we've come to the conclusion that the old way of doing it—tried and tested over the years—makes for good firewood. On top of that it fits in nicely with the rest of the work on the farm. For example, the work of splitting and stacking can be fitted in around the dairy farming and before we get down to the big spring chores."

"Maybe we could save ourselves a bit of time by using a wood processor," says Liv Kristin. "But it would probably be too much for what we need, and it would mean less hands-on contact with the wood. Anyway, we prefer to use moderate tools that aren't too rough on the wood. Machines that make a noise always used to irritate Peder's uncle. Whenever we bought some new tool for the tractor he would always say: 'Hm, I suppose this makes a humming sound too?' It was not because he was old-fashioned and irritable, but because he appreciated the silence you often find in manual work. We're the same," concludes Liv Kristin.

BURNING LOVE

I'm thinking about my neighbor again.

A year after he opened my eyes to the meaning of firewood, spring came around again, the way it always does. The same tractor came humming down the road, same as it did last year.

And Ottar was out there again, stacking his wood. Again the work seemed to make him stronger—a welcome reminder that, in spite of his ill health, he was still able to start and finish a large-scale piece of work.

A month later and the snow was all gone, and the wood stacked tidily away inside the shed. Ottar seemed in good spirits and still able to get about. But I did notice something. Something new in the way he looked at the piled wood.

Winter came.

One day an ambulance pulled up outside his house. There didn't seem to be any great hurry.

Perhaps that was what I had seen in his face when he closed the door to his woodshed on that spring morning: a realization that this particular woodpile would outlive him. That what he was actually doing was making things ready for his wife.

The woman he had been married to for more than fifty years.

I can see their house through the same kitchen window that I was looking through when I first thought of writing this book. From the white smoke drifting from the chimney I can see the wood is good and dry. His widow is warming herself by it now, as I lay down my pen.

OVERLEAF A classic view in the Norwegian countryside: firewood stacked and seasoned to perfection, covered by corrugated steel, held in place with old winter tires.

10
0
10
°C
20
20
30
30
40
40
50
50

COLD FACTS

UNITS OF MEASURE FOR WOOD

- 1 sm^3 = 1 solid cubic meter, that is, a volume of 1,000 liters of solid mass with no gaps. Often used for technical calculations.
- 1 lm^3 = 1 loose cubic meter, that is, a volume of 1,000 liters with gaps. Often used for stacked wood.
- 1 Norwegian cord of wood: 4 m x 1 m (or 2 m x 2 m) with wood cut into 60-cm lengths. A cord corresponds to 2.4 lm^3 stacked wood (loose mass) and 1.6 sm^3 (solid mass).
- 1 Norwegian greater cord of wood: 2 m x 2 m with wood cut into 3-meter lengths. Roughly corresponds to 12 lm^3 or between 6.7 and 7 sm^3.
- 1 US and Canadian cord of wood: 128 cubic feet of stacked wood, typically 4 x 8 x 4 feet (122 x 244 x 122 cm). Corresponds to 3.63 lm^3 stacked wood (loose mass) and 2.4 sm^3 (solid mass).
- Solid-mass percentage of stacked wood in 60-cm lengths is approximately 65 percent.*
- Solid-mass percentage of stacked wood in 30-cm lengths is approximately 74 percent.*
- Solid-mass percentage of loose wood (unstacked wood in a trailer or in a large sack) in 30-cm lengths is approximately 50 percent.*

* The percentages shown are the standard for Norwegian wood manufacturers. The basis for the calculation of stacked wood is that all wood will be more or less equally twisted and have branches and other irregularities that create openings in the layer of wood. Therefore, the solid mass will decrease the thinner and/or more finely chopped the logs are. Furthermore, the mass will decrease the *longer* the wood is. The reason for this is that a branch or bend will create an opening in the layer of wood equal to the length of the wood.

Rate of Drying for Birch

The table shows the moisture percentage of birch wood dried under various conditions in Hedmark in Norway, an area with low air humidity in springtime. Drying commenced at the end of May and was done outdoors, but under roof. Logs were an average of 14 inches (34 centimeters) long. The slightly increased moisture content in autumn is due to the increased moisture content of the air.

The figures indicate that wood dries surprisingly fast under optimum conditions. In humid climates, especially for very hard firewood like oak, two years of seasoning may be required. Longer logs will also require more drying time.

	Loosely stacked, good circulation, sun	Loosely stacked, some circulation, sun	Tightly stacked, some circulation, shade
Start	42.9%	42.9%	42.9%
2 days	32.8%	36.8%	40.7%
1 week	26.6%	29.8%	36.7%
2 weeks	23.7%	26.5%	32.8%
3 weeks	20.1%	23.7%	30%
4 weeks	18.5%	21.8%	28.7%
5 weeks	17.1%	20.1%	27.5%
2 months	16.4%	18.4%	24.7%
3 months	15%	16.8%	21.7%
6 months	16.9%	18.1%	20.6%

Heating Values

The table shows density and heating values for various trees in Norway. The density figures are average, derived from various forests around the country, and are typical for North European species. The density of some species, like spruce, may vary as much as 30 percent.

The figures in the table, including Btu per cord, applies to completely dry wood—that is, wood with a zero percent moisture content. To achieve real-life figures, they are intended to be recalculated according to the moisture content and stove efficiency in a given situation.

The reference figure for all wood species is that they provide 5.32 kilowatt-hours (kWh) for every 2.2 pounds (one kilogram) of completely dry wood. Factoring in a 20 percent moisture content, 2.2 pounds of "normally dry" wood will therefore yield 4.2 kilowatt-hours. Burned in a stove with 75 percent efficiency, the final energy output would be 3.2 kilowatt-hours for every kilogram.

Species	Weight per sm^3 (solid mass) in kilograms	kWh per sm^3 (solid mass)	Weight per cord (stacked mass in kilograms)	kWh per cord (stacked mass)	mBtu per lm^3 (stacked mass)	Weight lbs per American cord	mBtu per American cord
Holly	675	3,591	450	2,394	8.17	3,600	29.65
Hornbeam	660	3,511	440	2,341	7.99	3,520	28.99
Hickory	640	3,405	427	2,270	7.74	3,414	28.11
Black locust	570	3,032	380	2,022	6.90	3,040	25.04
Beech	570	3,032	380	2,022	6.90	3,040	25.04
Oak	550	2,926	367	1,951	6.66	2,934	24.16
Ash	550	2,926	367	1,951	6.66	2,934	24.16
Elm	540	2,873	360	1,915	6.53	2,880	23.72
Rowan	540	2,873	360	1,915	6.53	2,880	23.72
Tamarack	540	2,873	360	1,915	6.53	2,880	23.72
Maple	530	2,820	353	1,880	6.41	2,827	23.28
Douglas fir	520	2,766	347	1,844	6.29	2,774	22.84
Birch	500	2,660	333	1,773	6.05	2,667	21.96
Pine	440	2,341	293	1,561	5.32	2,347	19.33
Black alder	440	2,341	293	1,561	5.32	2,347	19.33
Willow	430	2,288	287	1,525	5.20	2,293	18.89
Linden	430	2,288	287	1,525	5.20	2,293	18.89
Aspen	400	2,128	267	1,419	4.84	2,133	17.57
Spruce	380	2,022	253	1,348	4.60	2,027	16.69
Poplar	380	2,022	253	1,348	4.60	2,027	16.69
Speckled alder	360	1,915	240	1,277	4.36	1,920	15.81
Western cedar	320	1,702	213	1,135	3.87	1,707	14.06

Source: Norwegian Forest and Landscape Institute, Ås

Ash Content as a Percentage of Dry Weight

	Birch	Pine	Spruce
Trunk	0.4	0.4	0.5
Bark	2.2	2.6	3.2
Branches	1.2	1	1.9
Sprigs or foliage	5.5	2.4	5.1
Whole tree with sprigs or foliage	1	0.9	1.6
Whole tree without sprigs or foliage	0.8	0.8	1.3

Volume of a Birch Tree

Diameter is measured four feet (1.3 meters) above the ground. Volume is indicated in liters.

Diameter in inches (centimeters)	Height in feet (meters)				
	37.80 (10)	49 (15)	65.5 (20)	80 (25)	100 (30)
4 (10)	38	54	69		
6 (15)	84	120	156	192	
8 (20)	142	206	271	336	401
10 (25)	209	309	410	512	614
12 (30)		426	571	717	865
16 (40)		686	944	1,204	1,464
20 (50)			1,364	1,768	2,174

Source: Norwegian Forest and Landscape Institute, Ås

Number of Kilowatt-Hours in a Birch Tree

These are calculated with a 20 percent moisture content and a stove with 75 percent efficiency.

Diameter in inches (centimeters)	Height in feet (meters)				
	37.80 (10)	49 (15)	65.5 (20)	80 (25)	100 (30)
4 (10)	76	108	138		
6 (15)	168	239	311	383	
8 (20)	283	411	541	670	800
10 (25)	417	616	818	1,021	1,225
12 (30)		850	1,139	1,430	1,726
16 (40)		1,369	1,883	2,402	2,921
20 (50)			2,721	3,527	4,337

Felling Requirements

The table shows the number of birch trees of equal size that must be felled to generate twelve thousand kilowatt-hours. The calculation is based on a 20 percent moisture content in the wood and a stove with 75 percent efficiency.

Diameter in inches (centimeters)	Height in feet (meters)				
	37.80 (10)	49 (15)	65.5 (20)	80 (25)	100 (30)
4 (10)	158	111	87		
6 (15)	72	50	39	31	
8 (20)	42	29	22	18	15
10 (25)	29	19	15	12	10
12 (30)		14	11	8	7
16 (40)		9	6	5	4
20 (50)			4	3.5	3

SOURCES

Chain Saws
Stihl / www.stihlusa.com
Jonsered / www.jonsered.com
Husqvarna / www.husqvarna.com
Partner / www.partner.biz/int/

Axes
Øyo Brothers / www.oeyo.no/home
Hultafors / www.hultafors.com
Gränsfors / www.gransforsbruk.com/en
Wetterlings / www.wetterlings.com
Fiskars / www3.fiskars.com
Leveraxe / www.vipukirves.fi

Closed Iron Stoves
Morsø / www.morsona.com
Jøtul / www.jotul.com
Contura / www.contura.eu

Soapstone Stoves
Granit Kleber / www.norskkleber.no/en/

Kitchen Stoves
Josef Davidssons Eftr / www.vedspis.se/en/
La Nordica / www.lanordica-extraflame.com/en/

Related Organizations
National Firewood Association / www.nationalfirewoodassociation.org
Chimney Safety Institute of America / www.csia.org
Don't Move Firewood / www.dontmovefirewood.org
Wood Heat / www.woodheat.org
Alliance for Green Heat / www.forgreenheat.org
Forestry Commission UK / www.forestry.gov.uk
Firewood Association of Australia / www.firewood.asn.au

REFERENCES

Børli, Hans. *Med øks og lyre: Blar av en tømmerhuggers dagbok*. Oslo: Aschehoug, 1988.
Cook, Dudley. *Keeping Warm with an Ax*. New York: Universe Books, 1981.
Hamran, Ulf. *Gamle ovner i Norge*. Oslo: C. Huitfeldt Forlag, 1989.
Herikstad, Per Erik. *Vedfyring og varme*. Oslo: Landbruksforlaget, 1995.
Hohle, Erik Eid, ed. *Bioenergi*. Brandbu, Norway: Energigården, 2001.
Langhammer, Aage, ed. *Bjørk, osp, or*. Ås, Norway: Norges Landbrukshøgskole, 1977.
Lindbekk, Bjarne. *Våre skogstrær*. Stavern, Norway: Omega forlag, 2000.
Logan, William Bryant. *Oak: The Frame of Civilization*. New York: W. W. Norton, 2005.
Moe, Daniel. *Svenska hushålls vedarbete*. Uppsala, Sweden: Sveriges Lantbruksuniversitet, 2007.
More, David, and John White. *Illustrated Trees of Britain and Northern Europe*. London: A. & C. Black, 2012.
Philbrick, Frank, and Stephen Philbrick. *The Backyard Lumberjack: The Ultimate Guide to Felling, Bucking, Splitting, and Stacking*. North Adams, MA: Storey, 2006.
Ryd, Yngve. *Eld*. Stockholm: Natur och Kultur, 2005.
Skogstad, Per, ed. *Treteknisk håndbok nr. 4*. Oslo: Treteknisk Institutt, 2009.
Solli, Svein. *Fyring med ved og brenselflis*. Oslo: Landbruksforlaget, 1980.
Thoreau, Henry David. *Walden; or, Life in the Woods*. Boston: Ticknor & Fields, 1854.
Vevstad, Andreas. Motorsaga. *Årbok for Norsk Skogbruksmuseum* 9 (1978–81).

ACKNOWLEDGMENTS

Thanks to Arne Fjeld, Ole Haugen, Ruben Knutsen, Liv Kristin Brenden, and Peder Brenden. And to Simen Gjølsjø at the Norwegian Forest and Landscape Institute, Edvard Karlsvik at SINTEF, Henning Horn at the Norwegian Institute of Wood Technology, Kristin Aasestad at the Central Bureau of Statistics of Norway, Magnar Eikerol at Gjøvik University College, and, finally, Gunnar Wilhelmsen. All shared valuable experience, data, and advice.

First published in the Norwegian language as *Hel Ved* by Kagge Forlag AS in 2011
First published in Great Britain in 2015 by

MacLehose Press
An imprint of Quercus Publishing Ltd
Carmelite House
50 Victoria Embankment
London EC4Y 0DZ

An Hachette UK company

Editor: David Cashion; Cover Design: John Gall; Design: Darilyn Lowe Carnes; Illustrations: Heesang Lee; Production: Anet Sirna-Bruder

All photos: Lars Mytting, except p. 2: Knut By/Tinagent; p. 22: Pekka Kyytinen/Helsinki City Museum; p. 50: Jonsered Sweden; p. 59: Stig Erik Tangen; p. 87: Roar Greipsland; p. 112: Christopher MacLehose; p. 113: Erling Gjøstøl; p. 119: Morten Aas; p. 139: Inge Hådem; p. 148: Granit Kleber AS; p. 154: Helge Løkamoen; p. 166: Ole Martin Mybakken

THIS TRANSLATION HAS BEEN PUBLISHED WITH THE FINANCIAL SUPPORT OF NORLA

A CIP catalogue record for this book is available from the British Library.

ISBN (HB) 978-0-85705-255-1
ISBN (Ebook) 978-1-78206-662-0

10 9 8

Printed and bound in Italy by L.E.G.O. S.p.A.